Why We Sleep

UNLOCKING THE POWER OF SLEEP AND DREAMS

Matthew Walker, PhD

SCRIBNER
New York London Toronto Sydney New Delhi

This publication contains the opinions and ideas of its author. It is intended to provide helpful and informative material on the subjects addressed in the publication. It is sold with the understanding that the author and publisher are not engaged in rendering medical, health, or any other kind of personal professional services in the book. The reader should consult his or her medical, health, or other competent professional before adopting any of the suggestions in the book or drawing inferences from it. The author and publisher specifically disclaim all responsibility for any liability, loss or risk, personal or otherwise, which is incurred as a consequence, directly or indirectly, of the use and application of any of the contents of this book.

Scribner
An Imprint of Simon & Schuster, Inc.
1230 Avenue of the Americas
New York, NY 10020

First Scribner trade paperback edition June 2018

SCRIBNER and design are registered trademarks of The Gale Group, Inc., used under license by Simon & Schuster, Inc., the publisher of this work.

For information about special discounts for bulk purchases, please contact Simon & Schuster Special Sales at 1-866-506-1949 or business@simonandschuster.com.

The Simon & Schuster Speakers Bureau can bring authors to your live event. For more information or to book an event contact the Simon & Schuster Speakers Bureau at 1-866-248-3049 or visit our website at www.simonspeakers.com.

Interior design by Jill Putorti

Manufactured in the United States of America

28

ISBN 978-1-5011-4431-8
ISBN 978-1-5011-4432-5 (pbk)
ISBN 978-1-5011-4433-2 (ebook)

Illustration credits appear on page 343.

To Dacher Keltner, for inspiring me to write.

Contents

– *Part 4* –

From Sleeping Pills to Society Transformed

PART 1

This Thing Called Sleep

CHAPTER 1

To Sleep . . .

Do you think you got enough sleep this past week? Can you recall the last time you woke up without an alarm clock feeling refreshed, not needing caffeine? If the answer to either of these questions is "no," you are not alone. More than a third of adults in many developed nations fail to obtain the recommended seven to nine hours of nightly sleep.*

I doubt you are surprised by this fact, but you may be surprised by the consequences. Routinely sleeping less than six hours a night weakens your immune system, substantially increasing your risk of certain forms of cancer. Insufficient sleep appears to be a key lifestyle factor linked to your risk of developing Alzheimer's disease. Inadequate sleep—even moderate reductions for just one week—disrupts blood sugar levels so profoundly that you would be classified as pre-diabetic. Short sleeping increases the likelihood of your coronary arteries becoming blocked and brittle, setting you on a path toward cardiovascular disease, stroke, and congestive heart failure. Fitting Charlotte Brontë's prophetic wisdom that "a ruffled mind makes a restless pillow," sleep disruption further contributes to all major psychiatric conditions, including depression, anxiety, and suicidality.

Perhaps you have also noticed a desire to eat more when you're tired? This is no coincidence. Too little sleep swells concentrations of a hormone that makes you feel hungry while suppressing a companion hormone that otherwise signals food satisfaction. Despite

*The Centers for Disease Control (CDC) stipulates that adults need seven hours of sleep or more per twenty-four hours.

being full, you still want to eat more. It's a proven recipe for weight gain in sleep-deficient adults and children alike. Worse, should you attempt to diet but don't get enough sleep while doing so, it is futile, since most of the weight you lose will come from lean body mass, not fat.

Add the above health consequences up, and a proven link becomes easier to accept: relative to the recommended seven to nine hours, the shorter your sleep, the shorter your life span. The old maxim "I'll sleep when I'm dead" is therefore unfortunate. Adopt this mind-set, and it is possible that you will be dead sooner and the quality of that (shorter) life will be worse. The elastic band of sleep deprivation can stretch only so far before it snaps. Sadly, human beings are in fact the only species that will deliberately deprive themselves of sleep without legitimate gain. Numerous components of wellness, and countless seams of societal fabric, are being eroded by our costly state of sleep neglect: human and financial alike. So much so that the Centers for Disease Control declared insufficient sleep as a public health epidemic. It may not be a coincidence that countries where sleep time has declined most dramatically over the past century, such as the US, the UK, Japan, and South Korea, and several in western Europe, are also those suffering the greatest increase in rates of the aforementioned physical diseases and mental disorders.

Scientists such as myself have even started lobbying doctors to start "prescribing" sleep. As medical advice goes, it's perhaps the most painless and enjoyable to follow. Do not, however, mistake this as a plea to doctors to start prescribing more sleeping *pills*—quite the opposite, in fact, considering the evidence surrounding the deleterious health consequences of these drugs.

But can we go so far as to say that a lack of sleep can kill you outright? Quite posssibly—on at least two counts. First, there is a very rare genetic disorder that starts with a progressive insomnia, emerging in midlife. Several months into the disease course, the patient stops sleeping altogether. By this stage, they have started to lose many basic brain and body functions. Few drugs that we currently have will help the patient sleep. After twelve to eighteen months of no sleep, the patient will die.

Second is the deadly circumstance of getting behind the wheel of a motor vehicle without having had sufficient sleep. Drowsy driving is the cause of hundreds of thousands of traffic accidents and fatalities each year. And here, it is not only the life of the sleep-deprived individuals that is at risk, but the lives of those around them. Tragically, one person dies in a traffic accident every hour in the United States due to a fatigue-related error.

Society's apathy toward sleep has, in part, been caused by the historic failure of science to explain why we need it. Sleep remained one of the last great biological mysteries. All of the mighty problem-solving methods in science—genetics, molecular biology, and high-powered digital technology—have been unable to unlock the stubborn vault of sleep. Minds of the most stringent kind, including Nobel Prize–winner Francis Crick, who deduced the twisted-ladder structure of DNA, famed Roman educator and rhetorician Quintilian, and even Sigmund Freud had all tried their hand at deciphering sleep's enigmatic code, all in vain.

To better frame this state of prior scientific ignorance, imagine the birth of your first child. At the hospital, the doctor enters the room and says, "Congratulations, it's a healthy baby boy. We've completed all of the preliminary tests and everything looks good." She smiles reassuringly and starts walking toward the door. However, before exiting the room she turns around and says, "There is just one thing. From this moment forth, and for the rest of your child's entire life, he will repeatedly and routinely lapse into a state of apparent coma. It might even resemble death at times. And while his body lies still his mind will often be filled with stunning, bizarre hallucinations. This state will consume one-third of his life and I have absolutely no idea why he'll do it, or what it is for. Good luck!"

Astonishing, but until very recently, this was reality: doctors and scientists could not give you a consistent or complete answer as to why we sleep. Consider that we have known the functions of the three other basic drives in life—to eat, to drink, and to reproduce—for many tens if not hundreds of years now. Yet the fourth main biological drive, common across the animal kingdom—the drive to sleep—has continued to elude science for millennia.

Addressing the question of why we sleep from an evolutionary perspective only compounds the mystery. No matter what vantage point you take, sleep would appear to be the most foolish of biological phenomena. When you are asleep, you cannot gather food. You cannot socialize. You cannot find a mate and reproduce. You cannot nurture or protect your offspring. Worse still, sleep leaves you vulnerable to predation. Sleep is surely one of the most puzzling of all human behaviors.

On any one of these grounds—never mind all of them in combination—there ought to have been a strong evolutionary pressure to *prevent* the emergence of sleep or anything remotely like it. As one sleep scientist has said, "If sleep does not serve an absolutely vital function, then it is the biggest mistake the evolutionary process has ever made."*

Yet sleep has persisted. Heroically so. Indeed, every animal species carefully studied to date sleeps.† This suggests that sleep evolved with—or very soon after—life itself on our planet. Moreover, the subsequent perseverance of sleep throughout evolution means there must be tremendous benefits that far outweigh all of the obvious hazards and detriments.

Ultimately, asking "Why do we sleep?" was the wrong question. It implied there was a single function, one holy grail of a reason that we slept, and we went in search of it. Theories ranged from the logical (a time for conserving energy), to the peculiar (an opportunity for eyeball oxygenation), to the psychoanalytic (a non-conscious state in which we fulfill repressed wishes).

This book will reveal a very different truth: sleep is infinitely more complex, profoundly more interesting, and strikingly health-relevant. We sleep for a rich litany of functions, plural—an abundant constellation of nighttime benefits that service both our brains and our bodies. There does not seem to be one major organ within the body, or process within the brain, that isn't optimally enhanced by sleep (and detrimentally impaired when we don't get enough). That we receive such a bounty of health benefits each night should not be surprising. After

---

*Dr. Allan Rechtschaffen.

†Cirelli, C., and Tononi, G. (2008). "Is sleep essential?" *PLoS Biology*. 6, e216.

all, we are *awake* for two-thirds of our lives, and we don't just achieve one useful thing during that stretch of time. We accomplish myriad undertakings that promote our own well-being and survival. Why, then, would we expect sleep—and the twenty-five to thirty years, on average, it takes from our lives—to offer one function only?

Through an explosion of discoveries over the past twenty years, we have come to realize that evolution did not make a spectacular blunder in conceiving of sleep. Sleep dispenses a multitude of health-ensuring benefits, yours to pick up in repeat prescription every twenty-four hours, should you choose.

Within the brain, sleep enriches a diversity of functions, including our ability to learn, memorize, and make logical decisions and choices. Benevolently servicing our psychological health, sleep recalibrates our emotional brain circuits, allowing us to navigate next-day social and psychological challenges with cool-headed composure. We are even beginning to understand the most impervious and controversial of all conscious experiences: the dream. Dreaming provides a unique suite of benefits to all species fortunate enough to experience it, humans included. Among these gifts are a consoling neurochemical bath that mollifies painful memories and a virtual reality space in which the brain melds past and present knowledge, inspiring creativity.

Downstairs in the body, sleep restocks the armory of our immune system, preventing infection and warding off all manner of sickness. Sleep reforms the body's metabolic state by fine-tuning the balance of insulin and circulating glucose. Sleep further regulates our appetite, helping control body weight through healthy food selection rather than rash impulsivity. Plentiful sleep maintains a flourishing microbiome within your gut from which we know so much of our nutritional health begins. Adequate sleep is intimately tied to the fitness of our cardiovascular system, lowering blood pressure while helping keep our hearts in fine condition.

A balanced diet and exercise are of vital importance, yes. But we now see sleep as a preeminent force in this health trinity. The physical and mental impairments caused by one night of bad sleep dwarf those caused by an equivalent absence of food or exercise. It is difficult to imagine any other state—natural or medically manipulated—that affords a more powerful redressing of physical and mental health at every level of analysis.

Based on a rich, new scientific understanding of sleep, we no longer have to ask what sleep is good for. Instead, we are now forced to wonder whether there are any biological functions that do *not* benefit by a good night's sleep.

Emerging from this research renaissance is an unequivocal message: sleep is the single most effective thing we can do to reset our brain and body health each day—Mother Nature's best effort yet at contra-death. Unfortunately, the real evidence that makes clear all of the dangers that befall individuals and societies when sleep becomes short have not been clearly telegraphed to the public. It is perhaps the most glaring omission in the contemporary health conversation. In response, this book is intended to help address this unmet need, and provide what I hope is a fascinating journey of discoveries. It aims to revise our cultural appreciation of sleep, and reverse our neglect of it.

Personally, I should note that I am rather in love with sleep (not just my own, though I do give myself a non-negotiable eight-hour sleep opportunity each night). I am in love with everything sleep is and does. I am in love with discovering all that remains unknown about it. I am in love with communicating the relevance of it to the public. I am in love with finding any and all methods for reuniting humanity with the sleep it so desperately needs. This love affair has now spanned a twenty-plus-year research career that began when I was a professor of psychiatry at Harvard Medical School and continues now that I am a professor of neuroscience and psychology at the University of California, Berkeley.

It was not, however, love at first sight. I am an accidental sleep researcher. It was never my intent to inhabit this esoteric outer territory of science. At age eighteen I went to study at the Queen's Medical Center in England: a prodigious institute in Nottingham boasting a wonderful band of brain scientists on its faculty. Ultimately, medicine wasn't for me, as it seemed more concerned with answers, whereas I was always more enthralled by questions. For me, answers were simply a way to get to the next question. I decided to study neuroscience, and after graduating, obtained my PhD in neurophysiology supported by a fellowship from England's Medical Research Council, London.

It was during my PhD work, conducted mostly at Newcastle University, that I began making my first scientific contributions in the field of sleep research. I was examining patterns of electrical brainwave activity in older adults in the early stages of dementia. Counter to common belief, there isn't just one type of dementia. Alzheimer's disease is the most common, but is only one of many types. For a number of treatment reasons, it is critical to know which type of dementia an individual is suffering from as soon as possible.

I began assessing brainwave activity from my patients during wake and sleep. My hypothesis: there was a unique and specific electrical brain signature that could forecast which dementia subtype each individual was progressing toward. Measurements taken during the day were ambiguous, with no clear signature of difference to be found. Only in the nighttime ocean of *sleeping* brainwaves did the recordings speak out a clear labeling of my patients' saddening disease fate. The discovery proved that sleep could potentially be used as a new early diagnostic litmus test to understand which type of dementia an individual would develop.

Sleep became my obsession. The answer it had provided me, like all good answers, only led to more fascinating questions, among them: Was the disruption of sleep in my patients actually contributing to the diseases they were suffering from, and even causing some of their terrible symptoms, such as memory loss, aggression, hallucinations, delusions? I read all I could. A scarcely believable truth began to emerge—nobody actually knew the clear reason why we needed sleep, and what it does. I could not answer my own question about dementia if this fundamental first question remained unanswered. I decided I would try to crack the code of sleep.

I halted my research in dementia and, for a post-doctoral position that took me across the Atlantic Ocean to Harvard, set about addressing one of the most enigmatic puzzles of humanity—one that had eluded some of the best scientists in history: Why do we sleep? With genuine naïveté, not hubris, I believed I would find the answer within two years. That was twenty years ago. Hard problems care little about what motivates their interrogators; they meter out their lessons of difficulty all the same.

Now, after two decades of my own research efforts, combined with thousands of studies from other laboratories around the world, we have many of the answers. These discoveries have taken me on wonderful, privileged, and unexpected journeys inside and outside of academia—from being a sleep consultant for the NBA, NFL, and British Premier League football teams; to Pixar Animation, government agencies, and well-known technology and financial companies. These sleep revelations, together with many similar discoveries from my fellow sleep scientists, will offer proof about the vital importance of sleep.

A final comment on the structure of this book. The chapters are written in a logical order, traversing a narrative arc in four main parts.

Part 1 demystifies this beguiling thing called sleep: what it is, what it isn't, who sleeps, how much they sleep, how human beings should sleep (but are not), and how sleep changes across your life span or that of your child, for better and for worse.

Part 2 details the good, the bad, and the deathly of sleep and sleep loss. We will explore all of the astonishing benefits of sleep for brain and for body, affirming what a remarkable Swiss Army knife of health and wellness sleep truly is. Then we turn to how and why a lack of sufficient sleep leads to a quagmire of ill health, disease, and untimely death—a wakeup call to sleep if ever there was one.

Part 3 offers safe passage from sleep to the fantastical world of dreams scientifically explained. From peering into the brains of dreaming individuals, and precisely how dreams inspire Nobel Prize–winning ideas that transform the world, to whether or not dream control really is possible, and if such a thing is even wise—all will be revealed.

Part 4 seats us first at the bedside, explaining numerous sleep disorders, including insomnia. I will unpack the obvious and not-so-obvious reasons for why so many of us find it difficult to get a good night's sleep, night after night. A discussion of sleeping pills then follows, based on scientific and clinical data. Details of new, safer, and more effective non-drug therapies for better sleep will then be described. Transitioning from bedside up to the level of sleep in society, we will subsequently learn of the sobering impact that insufficient sleep has in education, in medicine and health care, and in business. The evidence shatters beliefs about the usefulness of long waking hours with little sleep in effectively, safely, profitably, and ethically accomplishing the goals of each of these disciplines. Concluding the book with genuine optimistic hope, I lay out a road map of ideas that can reconnect humanity with the sleep it remains so bereft of—a new vision for sleep in the twenty-first century.

I should note that you need not read this book in this progressive, four-part narrative arc. Each chapter can, for the most part, be read individually, and out of order, without losing too much of its significance. I therefore invite you to consume the book in whole or in part, buffet-style or in order, all according to your personal taste.

It is worthwhile pointing out that this book is not designed to be a self-help guide. It is not written to target or treat sleep disorders, including insomnia. There are books that do this, and many of them will recommend speaking to a doctor if you suspect you have a sleep disorder. I am also very understanding of, and sympathetic to, those people who struggle with sleep and are most anxious about it. For these individuals, it is possible that their anxiety may increase when reading about the impact of insufficient sleep, including information contained in the book. I therefore want to alert the reader to this possibility, allowing for reader discretion on this matter.

In closing, I offer a disclaimer. Should you feel drowsy and fall asleep while reading the book, unlike most authors, I will not be disheartened. Indeed, based on the topic and content of this book, I am actively going to encourage that kind of behavior from you. Knowing what I know about the relationship between sleep and memory, it is the greatest form of flattery for me to know that you, the reader, cannot resist the urge to strengthen and thus remember what I am telling you by falling asleep. So please, feel free to ebb and flow into and out of consciousness during this entire book. I will take absolutely no offense. On the contrary, I would be delighted.

CHAPTER 2

Caffeine, Jet Lag, and Melatonin

Losing and Gaining Control of Your Sleep Rhythm

How does your body know when it's time to sleep? Why do you suffer from jet lag after arriving in a new time zone? How do you overcome jet lag? Why does that acclimatization cause you yet more jet lag upon returning home? Why do some people use melatonin to combat these issues? Why (and how) does a cup of coffee keep you awake? Perhaps most importantly, how do you know if you're getting enough sleep?

There are two main factors that determine when you want to sleep and when you want to be awake. As you read these very words, both factors are powerfully influencing your mind and body. The first factor is a signal beamed out from your internal twenty-four-hour clock located deep within your brain. The clock creates a cycling, day-night rhythm that makes you feel tired or alert at regular times of night and day, respectively. The second factor is a chemical substance that builds up in your brain and creates a "sleep pressure." The longer you've been awake, the more that chemical sleep pressure accumulates, and consequentially, the sleepier you feel. It is the balance between these two factors that dictates how alert and attentive you are during the day, when you will feel tired and ready for bed at night, and, in part, how well you will sleep.

GOT RHYTHM?

Central to many of the questions in the opening paragraph is the powerful sculpting force of your twenty-four-hour rhythm, also known as your circadian rhythm. Everyone generates a circadian rhythm (*circa*, meaning "around," and *dian*, derivative of *diam*, meaning "day"). Indeed, most living creatures on the planet with a life span of more than several days generate this natural cycle. The internal twenty-four-hour clock within your brain communicates its daily circadian rhythm signal to every other region of your brain and every organ in your body.

Your twenty-four-hour tempo helps to determine when you want to be awake and when you want to be asleep. But it controls other rhythmic patterns, too. These include your timed preferences for eating and drinking, your moods and emotions, the amount of urine you produce,* your core body temperature, your metabolic rate, and the release of numerous hormones. It is no coincidence that the likelihood of breaking an Olympic record has been clearly tied to time of day, being maximal at the natural peak of the human circadian rhythm in the early afternoon. Even the timing of births and deaths demonstrates circadian rhythmicity due to the marked swings in key life-dependent metabolic, cardiovascular, temperature, and hormonal processes that this pacemaker controls.

Long before we discovered this biological pacemaker, an ingenious experiment did something utterly remarkable: stopped time—at least, for a plant. It was in 1729 when French geophysicist Jean-Jacques d'Ortous de Mairan discovered the very first evidence that plants generate their own internal time.

De Mairan was studying the leaf movements of a species that displayed heliotropism: when a plant's leaves or flowers track the trajectory of the sun as it moves across the sky during the day. De Mairan was intrigued by one plant in particular, called *Mimosa pudica*.† Not only do the leaves of this plant trace the arching daytime passage of the sun across the sky's face, but at night, they collapse down, almost as though they had wilted. Then, at the start of the following day, the leaves pop

*I should note, from personal experience, that this is a winning fact to dispense at dinner parties, family gatherings, or other such social occasions. It will almost guarantee nobody will approach or speak to you again for the rest of the evening, and you'll also never be invited back.

†The word *pudica* is from the Latin meaning "shy" or "bashful," since the leaves will also collapse down if you touch or stroke them.

open once again like an umbrella, healthy as ever. This behavior repeats each and every morning and evening, and it caused the famous evolutionary biologist Charles Darwin to call them "sleeping leaves."

Prior to de Mairan's experiment, many believed that the expanding and retracting behavior of the plant was solely determined by the corresponding rising and setting of the sun. It was a logical assumption: daylight (even on cloudy days) triggered the leaves to open wide, while ensuing darkness instructed the leaves to shut up shop, close for business, and fold away. That assumption was shattered by de Mairan. First, he took the plant and placed it out in the open, exposing it to the signals of light and dark associated with day and night. As expected, the leaves expanded during the light of day and retracted with the dark of night.

Then came the genius twist. De Mairan placed the plant in a sealed box for the next twenty-four-hour period, plunging it into total dark for both day and night. During these twenty-four hours of blackness, he would occasionally take a peek at the plant in controlled darkness, observing the state of the leaves. Despite being cut off from the influence of light during the day, the plant still behaved as though it were being bathed in sunlight; its leaves were proudly expanded. Then, it retracted its leaves as if on cue at the end of the day, even without the sun's setting signal, and they stayed collapsed throughout the entire night.

It was a revolutionary discovery: de Mairan had shown that a living organism kept its own time, and was not, in fact, slave to the sun's rhythmic commands. Somewhere within the plant was a twenty-four-hour rhythm generator that could track time without any cues from the outside world, such as daylight. The plant didn't just have a circadian rhythm, it had an "endogenous," or self-generated, rhythm. It is much like your heart drumming out its own self-generating beat. The difference is simply that your heart's pacemaker rhythm is far faster, usually beating at least once a second, rather than once every twenty-four-hour period like the circadian clock.

Surprisingly, it took another two hundred years to prove that we humans have a similar, internally generated circadian rhythm. But this experiment added something rather unexpected to our understanding of internal timekeeping. It was 1938, and Professor Nathaniel Kleitman at the University of Chicago, accompanied by his research assistant Bruce Richardson, were to perform an even more radical scientific study. It required a type of dedication that is arguably without match or comparison to this day.

Kleitman and Richardson were to be their own experimental guinea pigs. Loaded with food and water for six weeks and a pair of dismantled, high-standing hospital beds, they took a trip into Mammoth Cave in Kentucky, one of the deepest caverns on the planet—so deep, in fact, that no detectable sunlight penetrates its farthest reaches. It was from this darkness that Kleitman and Richardson were to illuminate a striking scientific finding that would define our biological rhythm as being *approximately* one day (circadian), and not *precisely* one day.

In addition to food and water, the two men brought a host of measuring devices to assess their body temperatures, as well as their waking and sleeping rhythms. This recording area formed the heart of their living space, flanked either side by their beds. The tall bed legs were each seated in a bucket of water, castle-moat style, to discourage the innumerable small (and not so small) creatures lurking in the depths of Mammoth Cave from joining them in bed.

The experimental question facing Kleitman and Richardson was simple: When cut off from the daily cycle of light and dark, would their biological rhythms of sleep and wakefulness, together with body temperature, become completely erratic, or would they stay the same as those individuals in the outside world exposed to rhythmic daylight? In total, they lasted thirty-two days in complete darkness. Not only did they aggregate some impressive facial hair, but they made two groundbreaking discoveries in the process. The first was that humans, like de Mairan's heliotrope plants, generated their own endogenous circadian rhythm in the absence of external light from the sun. That is, neither Kleitman nor Richardson descended into random spurts of wake and sleep, but instead expressed a predictable and repeating pattern of prolonged wakefulness (about fifteen hours), paired with consolidated bouts of about nine hours of sleep.

The second unexpected—and more profound—result was that their reliably repeating cycles of wake and sleep were not precisely twenty-four hours in length, but consistently and undeniably longer than twenty-four hours. Richardson, in his twenties, developed a sleep-wake cycle of between twenty-six and twenty-eight hours in length. That of Kleitman, in his forties, was a little closer to, but still longer than, twenty-four hours. Therefore, when removed from the external influence of daylight, the internally generated "day" of each man was not exactly twenty-four hours, but a little more than that. Like an inaccurate wristwatch whose time runs long, with each passing (real) day in the outside world, Kleitman and Richardson began to add time based on their longer, internally generated chronometry.

Since our innate biological rhythm is not precisely twenty-four hours, but thereabouts, a new nomenclature was required: the *circa*dian rhythm—that is, one that is *approximately*, or around, one day in length, and not precisely one day.* In the seventy-plus years since Kleitman and Richardson's seminal experiment, we have now determined that the average duration of a human adult's endogenous circadian clock runs around twenty-four hours and fifteen minutes in length. Not too far off the twenty-four-hour rotation of the Earth, but not the precise timing that any self-respecting Swiss watchmaker would ever accept.

Thankfully, most of us don't live in Mammoth Cave, or the constant darkness it imposes. We routinely experience light from the sun that comes to the rescue of our imprecise, overrunning internal circadian clock. Sunlight acts like a manipulating finger and thumb on the side-dial of an imprecise wristwatch. The light of the sun methodically resets our inaccurate internal timepiece each and every day, "winding" us back to precisely, not approximately, twenty-four hours.†

It is no coincidence that the brain uses daylight for this resetting purpose. Daylight is the most reliable, repeating signal that we have in our environment. Since the birth of our planet, and every single day thereafter without fail, the sun has always risen in the morning and set in the evening. Indeed, the reason most living species likely

*This phenomenon of an imprecise internal biological clock has now been consistently observed in many different species. However, it is not consistently long in all species, as it is in humans. For some, the endogenous circadian rhythm runs short, being less than twenty-four hours when placed in total darkness, such as hamsters or squirrels. For others, such as humans, it is longer than twenty-four hours.

†Even sunlight coming through thick cloud on a rainy day is powerful enough to help reset our biological clocks.

adopted a circadian rhythm is to synchronize themselves and their activities, both internal (e.g., temperature) and external (e.g., feeding), with the daily orbital mechanics of planet Earth spinning on its axis, resulting in regular phases of light (sun facing) and dark (sun hiding).

Yet daylight isn't the only signal that the brain can latch on to for the purpose of biological clock resetting, though it is the principal and preferential signal, when present. So long as they are reliably repeating, the brain can also use other external cues, such as food, exercise, temperature fluctuations, and even regularly timed social interaction. All of these events have the ability to reset the biological clock, allowing it to strike a precise twenty-four-hour note. It is the reason that individuals with certain forms of blindness do not entirely lose their circadian rhythm. Despite not receiving light cues due to their blindness, other phenomena act as their resetting triggers. Any signal that the brain uses for the purpose of clock resetting is termed a zeitgeber, from the German "time giver" or "synchronizer." Thus, while light is the most reliable and thus the primary zeitgeber, there are many factors that can be used in addition to, or in the absence of, daylight.

The twenty-four-hour biological clock sitting in the middle of your brain is called the suprachiasmatic (pronounced *soo-pra-kai-as-MAT-ik*) nucleus. As with much of anatomical language, the name, while far from easy to pronounce, is instructional: *supra*, meaning above, and *chiasm*, meaning a crossing point. The crossing point is that of the optic nerves coming from your eyeballs. Those nerves meet in the middle of your brain, and then effectively switch sides. The suprachiasmatic nucleus is located just above this intersection for a good reason. It "samples" the light signal being sent from each eye along the optic nerves as they head toward the back of the brain for visual processing. The suprachiasmatic nucleus uses this reliable light information to reset its inherent time inaccuracy to a crisp twenty-four-hour cycle, preventing any drift.

When I tell you that the suprachiasmatic nucleus is composed of 20,000 brain cells, or neurons, you might assume it is enormous, consuming a vast amount of your cranial space, but actually it is tiny. The brain is composed of approximately 100 billion neurons, making the suprachiasmatic nucleus minuscule in the relative scheme of cerebral matter. Yet despite its stature, the influence of the suprachiasmatic nucleus on the rest of the brain and the body is anything but meek. This tiny clock is the central conductor of life's biological rhythmic symphony—yours and every other living species. The suprachiasmatic nucleus controls a vast array of behaviors, including our focus in this chapter: when you want to be awake and asleep.

For diurnal species that are active during the day, such as humans, the circadian rhythm activates many brain and body mechanisms in the brain and body during daylight hours that are designed to keep you awake and alert. These processes are then ratcheted down at nighttime, removing that alerting influence. Figure 1 shows one such example of a circadian rhythm—that of your body temperature. It represents average core body temperature (rectal, no less) of a group of human adults. Starting at "12 pm" on the far left, body temperature begins to rise, peaking late in the afternoon. The trajectory then changes. Temperature begins to decline again, dropping below that of the midday start-point as bedtime approaches.

Figure 1: Typical Twenty-Four-Hour Circadian Rhythm (Core Body Temperature)

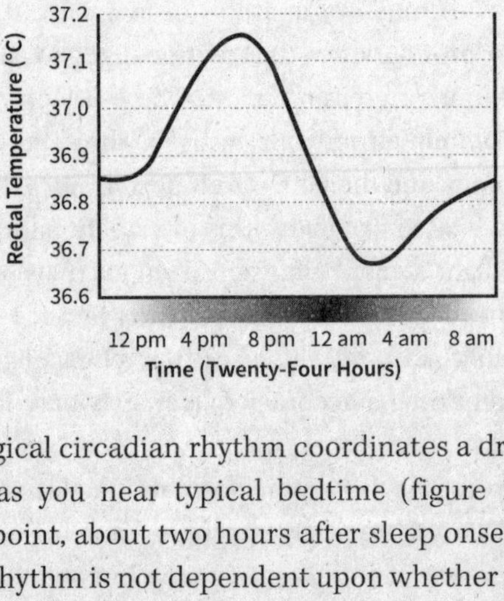

Your biological circadian rhythm coordinates a drop in core body temperature as you near typical bedtime (figure 1), reaching its nadir, or low point, about two hours after sleep onset. However, this temperature rhythm is not dependent upon whether you are actually asleep. If I were to keep you awake all night, your core body temperature would still show the same pattern. Although the temperature drop helps to initiate sleep, the temperature change itself will rise and fall across the twenty-four-hour period regardless of whether you are awake or asleep. It is a classic demonstration of a preprogrammed circadian rhythm that will repeat over and over without fail, like a metronome. Temperature is just one of many twenty-four-hour rhythms that the suprachiasmatic nucleus governs. Wakefulness and sleep are another. Wakefulness and sleep are therefore under the control of the circadian rhythm, and not the other way around. That is, your circadian rhythm will march up and down every twenty-four hours irrespective of whether you have slept or not. Your circadian rhythm is unwavering in this regard. But look across individuals, and you discover that not everyone's circadian timing is the same.

MY RHYTHM IS NOT YOUR RHYTHM

Although human beings display an unyielding twenty-four-hour pattern, the respective peak and trough points are strikingly different from one individual to the next. For some people, their peak of wakefulness arrives early in the day, and their sleepiness trough arrives early at night. These are "morning types," and make up about 40 percent of the populace. They prefer to wake at or around dawn, are happy to do so, and function optimally at this time of day. Others are "evening types," and account for approximately 30 percent of the population. They naturally prefer going to bed late and subsequently wake up late the following morning, or even in the afternoon. The remaining 30 percent of people lie somewhere in between morning and evening types, with a slight leaning toward eveningness, like myself.

You may colloquially know these two types of people as "morning larks" and "night owls," respectively. Unlike morning larks, night owls are frequently incapable of falling asleep early at night, no matter how hard they try. It is only in the early-morning hours that owls can drift off. Having not fallen asleep until late, owls of course strongly dislike waking up early. They are unable to function well at this time, one cause of which is that, despite being "awake," their brain remains in a more sleep-like state throughout the early morning. This is especially true of a region called the prefrontal cortex, which sits above the eyes, and can be thought of as the head office of the brain. The prefrontal cortex controls high-level thought and logical reasoning, and helps keep our emotions in check. When a night owl is forced to wake up too early, their prefrontal cortex remains in a disabled, "offline" state. Like a cold engine after an early-morning start, it takes a long time before it warms up to operating temperature, and before that will not function efficiently.

An adult's owlness or larkness, also known as their chronotype, is strongly determined by genetics. If you are a night owl, it's likely that one (or both) of your parents is a night owl. Sadly, society treats night owls rather unfairly on two counts. First is the label of being lazy, based on a night owl's wont to wake up later in the day, due to the fact that they did not fall asleep until the early-morning hours. Others (usually morning larks) will chastise night owls on the erroneous assumption that such preferences are a choice, and if they were not so slovenly, they could easily wake up early. However, night owls are not owls by choice. They are bound to a delayed schedule by unavoidable DNA hardwiring. It is not their *conscious* fault, but rather their *genetic* fate.

Second is the engrained, un-level playing field of society's work scheduling, which is strongly biased toward early start times that punish owls and favor larks. Although the situation is improving, standard employment schedules force owls into an unnatural sleep-wake rhythm. Consequently, job performance of owls as a whole is far less optimal in the mornings, and they are further prevented from expressing their true performance potential in the late afternoon and early evening as standard work hours end prior to its arrival. Most unfortunately, owls are more chronically sleep-deprived, having to wake up with the larks, but not being able to fall asleep until far later in the evening. Owls are thus often forced to burn the proverbial candle at both ends. Greater ill health caused by a lack of sleep therefore befalls owls, including higher rates of depression, anxiety, diabetes, cancer, heart attack, and stroke.

In this regard, a societal change is needed, offering accommodations not dissimilar to those we make for other physically determined differences (e.g., sight impaired). We require more supple work schedules that better adapt to all chronotypes, and not just one in its extreme.

You may be wondering why Mother Nature would program this variability across people. As a social species, should we not all be synchronized and therefore awake at the same time to promote maximal human interactions? Perhaps not. As we'll discover later in this book, humans likely evolved to co-sleep as families or even whole tribes, not alone or as couples. Appreciating this evolutionary context, the benefits of such genetically programmed variation in sleep/wake timing preferences can be understood. The night owls in the group would not be going to sleep until one or two a.m., and not waking until nine or ten a.m. The morning larks, on the other hand, would have retired for the night at nine p.m. and woken at five a.m. Consequently, the group as a whole is only collectively vulnerable (i.e., every person asleep) for just four rather than eight hours, despite everyone still getting the chance for eight hours of sleep. That's potentially a 50 percent increase in survival fitness. Mother Nature would never pass on a biological trait—here, the useful variability in when individuals within a collective tribe go to sleep and wake up—that could enhance the survival safety and thus fitness of a species by this amount. And so she hasn't.

MELATONIN

Your suprachiasmatic nucleus communicates its repeating signal of night and day to your brain and body using a circulating messenger called melatonin. Melatonin has other names, too. These include "the hormone of darkness" and "the vampire hormone." Not because it is sinister, but simply because melatonin is released at night. Instructed by the suprachiasmatic nucleus, the rise in melatonin begins soon after dusk, being released into the bloodstream from the pineal gland, an area situated deep in the back of your brain. Melatonin acts like a powerful bullhorn, shouting out a clear message to the brain and body: "It's dark, it's dark!" At this moment, we have been served a writ of nightime, and with it, a biological command for the timing of sleep onset.*

In this way, melatonin helps regulate the *timing* of when sleep occurs by systemically signaling darkness throughout the organism. But melatonin has little influence on the *generation* of sleep itself: a mistaken assumption that many people hold. To make clear this distinction, think of sleep as the Olympic 100-meter race. Melatonin is the voice of the timing official that says "Runners, on your mark," and then fires the starting pistol that triggers the race. That *timing* official (melatonin) governs when the race (sleep) begins, but does not participate in the race. In this analogy, the sprinters themselves are other brain regions and processes that actively *generate* sleep. Melatonin corrals these sleep-generating regions of the brain to the starting line of bedtime. Melatonin simply provides the official instruction to commence the event of sleep, but does not participate in the sleep race itself.

For these reasons, melatonin is not a powerful sleeping aid in and of itself, at least not for healthy, non-jet-lagged individuals (we'll explore jet lag—and how melatonin can be helpful—in a moment). There may be little, if any, quality melatonin in the pill. That said, there is a significant sleep placebo effect of melatonin, which should not be underestimated: the placebo effect is, after all, the most reliable effect in all of pharmacology. Equally important to realize is the fact that over-the-counter melatonin is not commonly regulated by governing bodies around the world, such as the US Food and Drug Administration (FDA). Scientific evaluations of over-the-counter brands have found melatonin concentrations that range from 83 percent less than that claimed on the label, to 478 percent more than that stated.†

Once sleep is under way, melatonin slowly decreases in concentration across the night and into the morning hours. With dawn, as sunlight enters the brain through the eyes (even through the closed lids), a brake pedal is applied to the pineal gland, thereby shutting off the release of melatonin. The absence of circulating melatonin now informs the brain

*For nocturnal species like bats, crickets, fireflies, or foxes, this call happens in the morning.

† L. A. Erland and P. K. Saxena, "Melatonin natural health products and supplements: presence of serotonin and significant variability of melatonin content," *Journal of Clinical Sleep Medicine* 2017; 13(2): 275–81.

and body that the finish line of sleep has been reached. It is time to call the race of sleep over and allow active wakefulness to return for the rest of the day. In this regard, we human beings are "solar powered." Then, as light fades, so, too, does the solar brake pedal blocking melatonin. As melatonin rises, another phase of darkness is signaled and another sleep event is called to the starting line.

You can see a typical profile of melatonin release in figure 2. It starts a few hours after dusk. Then it rapidly rises, peaking around four a.m. Thereafter, it begins to drop as dawn approaches, falling to levels that are undetectable by early to midmorning.

Figure 2: The Cycle of Melatonin

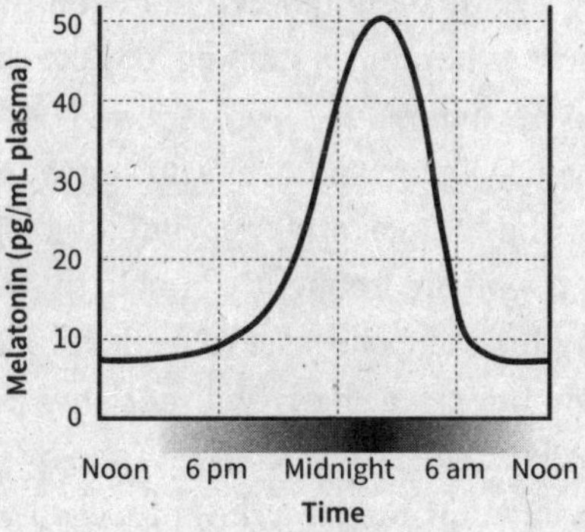

HAVE RHYTHM, WON'T TRAVEL

The advent of the jet engine was a revolution for the mass transit of human beings around the planet. However, it created an unforeseen biological calamity: jet planes offered the ability to speed through time zones faster than our twenty-four-hour internal clocks could ever keep up with or adjust to. Those jets caused a biological time lag: jet lag. As a result, we feel tired and sleepy during the day in a distant time zone because our internal clock still thinks it is nighttime. It hasn't yet caught up. If that were not bad enough, at night, we are frequently unable to initiate or maintain sleep because our internal clock now believes it to be daytime.

Take the example of my recent flight home to England from San Francisco. London is eight hours ahead of San Francisco. When I arrive in England, despite the digital clock in London's Heathrow Airport telling me it is nine a.m., my internal circadian clock is registering a very different time—California time, which is one a.m. I should be fast asleep. I will drag my time-lagged brain and body through the London day in a state of deep lethargy. Every aspect of my biology is demanding sleep; sleep that most people back in California are being swaddled in at this time.

The worst, however, is yet to come. By midnight London time, I am in bed, tired and wanting to fall asleep. But unlike most people in London, I can't seem to drift off. Though it is midnight in London, my internal biological clock believes it to be four p.m., which it is in California. I would normally be wide awake, and so I am, lying in bed in London. It will be five or six hours before my natural tendency to fall asleep arrives . . . just as London is starting to wake up, and I have to give a public lecture. What a mess.

This is jet lag: you feel tired and sleepy during the day in the new time zone because your body clock and associated biology still "think" it is nighttime. At night, you are frequently unable to sleep solidly because your biological rhythm still believes it to be daytime.

Fortunately, my brain and body will not stay in this mismatched limbo forever. I will acclimatize to London time by way of the sunlight signals in the new location. But it's a slow process. For every day you are in a different time zone, your suprachiasmatic nucleus can only readjust by about one hour. It therefore took me about eight days to readjust to London time after having been in San Francisco, since London is eight hours ahead of San Francisco. Sadly, after such epic efforts by my suprachiasmatic nucleus's twenty-four-hour clock to drag itself forward in time and get settled in London, it faces some depressing news: I now have to fly back to San Francisco after nine days. My poor biological clock has to suffer this struggle all over again in the reverse direction!

You may have noticed that it feels harder to acclimate to a new time zone when traveling eastward than when flying westward. There are at least two reasons for this. First, the eastward direction requires that you fall asleep earlier than you would normally, which is a tall biological order for the mind to simply will into action. In contrast, the westward direction requires you to stay up later, which is a consciously and pragmatically easier prospect. Second, you will remember that when shut off from any outside world influences, our natural circadian rhythm is innately longer than one day—about twenty-four hours and fifteen minutes. Modest as this may be, this makes it somewhat easier for you to artificially stretch a day than shrink it. When you travel westward—in the direction of your innately longer internal clock—that "day" is longer than twenty-four hours for you and why it feels a little easier to accommodate to. Eastward travel, however, which involves a "day" that is shorter in length for you than twenty-four hours, goes against the grain of your innately long internal rhythm to start with, which is why it is rather harder to do.

West or east, jet lag still places a torturous physiological strain on the brain, and a deep biological stress upon the cells, organs, and major systems of the body. And there are consequences. Scientists have studied airplane cabin crews who frequently fly on long-haul routes and have little chance to recover. Two alarming results have emerged. First, parts of their brains—specifically those related to learning and memory—had physically shrunk, suggesting the destruction of brain cells caused by the biological stress of time-zone travel. Second, their short-term memory was significantly impaired. They were considerably more forgetful than individuals of similar age and background who did not frequently travel through time zones. Other studies of pilots, cabin crew members, and shift workers have reported additionally disquieting consequences, including far higher rates of cancer and type 2 diabetes than the general population—or even carefully controlled match individuals who do not travel as much.

Based on these deleterious effects, you can appreciate why some people faced with frequent jet lag, including airline pilots and cabin crew, would want to limit such misery. Often, they choose to take melatonin pills in an attempt to help with the problem. Recall my flight from San Francisco to London. After arriving that day, I had real difficulty getting to sleep and staying asleep that night. In part, this was because melatonin was not being released during my nighttime in London. My melatonin rise was still many hours away, back on California time. But let's imagine that I was going to use a legitimate compound of melatonin after arriving in London. Here's how it works: at around seven to eight p.m. London time I would take a melatonin pill, triggering an artificial rise in circulating melatonin that mimics the natural melatonin spike currently occurring in most of the people in London. As a consequence, my brain is fooled into believing it's nighttime, and with that chemically induced trick comes the signaled timing of the sleep race. It will still be a struggle to generate the event of sleep itself at this irregular time (for me), but the timing signal does significantly increase the likelihood of sleep in this jet-lagged context.

SLEEP PRESSURE AND CAFFEINE

Your twenty-four-hour circadian rhythm is the first of the two factors determining wake and sleep. The second is sleep pressure. At this very moment, a chemical called adenosine is building up in your brain. It will continue to increase in concentration with every waking minute that elapses. The longer you are awake, the more adenosine will accumulate. Think of adenosine as a chemical barometer that continuously registers the amount of elapsed time since you woke up this morning.

One consequence of increasing adenosine in the brain is an increasing desire to sleep. This is known as sleep pressure, and it is the second force that will determine when you feel sleepy, and thus should go to bed. Using a clever dual-action effect, high concentrations of adenosine simultaneously turn down the "volume" of wake-promoting regions in the brain and turn up the dial on sleep-inducing regions. As a result of that chemical sleep pressure, when adenosine concentrations peak, an irresistible urge for slumber will take hold.* It happens to most people after twelve to sixteen hours of being awake.

You can, however, artificially mute the sleep signal of adenosine by using a chemical that makes you feel more alert and awake: caffeine. Caffeine is not a food supplement. Rather, caffeine is the most widely used (and abused) psychoactive stimulant in the world. It is the second most traded commodity on the planet, after oil. The consumption of caffeine represents one of the longest and largest unsupervised drug

*Assuming you have a stable circadian rhythm, and have not recently experienced jet travel through numerous time zones, in which case you can still have difficulty falling asleep even if you have been awake for sixteen hours.

studies ever conducted on the human race, perhaps rivaled only by alcohol, and it continues to this day.

Caffeine works by successfully battling with adenosine for the privilege of latching on to adenosine welcome sites—or receptors—in the brain. Once caffeine occupies these receptors, however, it does not stimulate them like adenosine, making you sleepy. Rather, caffeine blocks and effectively inactivates the receptors, acting as a masking agent. It's the equivalent of sticking your fingers in your ears to shut out a sound. By hijacking and occupying these receptors, caffeine blocks the sleepiness signal normally communicated to the brain by adenosine. The upshot: caffeine tricks you into feeling alert and awake, despite the high levels of adenosine that would otherwise seduce you into sleep.

Levels of circulating caffeine peak approximately thirty minutes after oral administration. What is problematic, though, is the persistence of caffeine in your system. In pharmacology, we use the term "half-life" when discussing a drug's efficacy. This simply refers to the length of time it takes for the body to remove 50 percent of a drug's concentration. Caffeine has an average half-life of five to seven hours. Let's say that you have a cup of coffee after your evening dinner, around 7:30 p.m. This means that by 1:30 a.m., 50 percent of that caffeine may still be active and circulating throughout your brain tissue. In other words, by 1:30 a.m., you're only halfway to completing the job of cleansing your brain of the caffeine you drank after dinner.

There's nothing benign about that 50 percent mark, either. Half a shot of caffeine is still plenty powerful, and much more decomposition work lies ahead throughout the night before caffeine disappears. Sleep will not come easily or be smooth throughout the night as your brain continues its battle against the opposing force of caffeine. Most people do not realize how long it takes to overcome a single dose of caffeine, and therefore fail to make the link between the bad night of sleep we wake from in the morning and the cup of coffee we had ten hours earlier with dinner.

Caffeine—which is not only prevalent in coffee, certain teas, and many energy drinks, but also foods such as dark chocolate and ice cream, as well as drugs such as weight-loss pills and pain relievers—is a common culprit that keeps people from falling asleep easily and sleeping soundly thereafter, typically masquerading as insomnia, an actual medical condition. Also be aware that *de-caffeinated* does not mean *non-caffeinated*. Depending on the decaffeination method and the bean that is used, one cup of decaf can have between 3 to as high as 10 percent of the dose of a regular cup of coffee.

The "jolt" of caffeine does wear off. Caffeine is removed from your system by an enzyme within your liver,* which gradually degrades it over time. Based in large part on genetics,† some people have a more efficient version of the enzyme that degrades caffeine, allowing the liver to rapidly clear it from the bloodstream. Others, however, have a slower-acting version of the enzyme. It takes far longer for their system to eliminate the same amount of caffeine. As a result, they are very sensitive to caffeine's effects. One cup of tea or coffee in the morning will last much of the day, and should they have a second cup, even early in the afternoon, they will find it difficult to fall asleep in the evening. Aging also alters the speed of caffeine clearance: the older we are, the longer it takes our brain and body to remove caffeine, and thus the more sensitive we become in later life to caffeine's sleep-disrupting influence.

If you are trying to stay awake late into the night by drinking coffee, you should be prepared for a nasty consequence when your liver successfully evicts the caffeine from your system: a phenomenon commonly known as a "caffeine crash." Like the batteries running down on a toy robot, your energy levels plummet rapidly. You find it difficult to function and concentrate, with a strong sense of sleepiness once again.

We now understand why. For the entire time that caffeine is in your system, the sleepiness chemical it blocks (adenosine) nevertheless continues to build up. Your brain is not aware of this rising tide of sleep-encouraging adenosine, however, because the wall of caffeine you've created is holding it back from your perception. But once your liver dismantles that barricade of caffeine, you feel a vicious backlash: you are hit with the sleepiness

---

*There are other factors that contribute to caffeine sensitivity, such as age, other medications currently being taken, and the quantity and quality of prior sleep. A. Yang, A. A. Palmer, and H. de Wit, "Genetics of caffeine consumption and responses to caffeine," *Psychopharmacology* 311, no. 3 (2010): 245–57, http://www.ncbi.nlm.nih.gov/pmc/articles/PMC4242593/.

†The principal liver enzyme that metabolizes caffeine is called cytochrome P450 1A2.

you had experienced two or three hours ago before you drank that cup of coffee *plus* all the extra adenosine that has accumulated in the hours in between, impatiently waiting for caffeine to leave. When the receptors become vacant by way of caffeine decomposition, adenosine rushes back in and smothers the receptors. When this happens, you are assaulted with a most forceful adenosine-trigger urge to sleep—the aforementioned caffeine crash. Unless you consume even more caffeine to push back against the weight of adenosine, which would start a dependency cycle, you are going to find it very, very difficult to remain awake.

To impress upon you the effects of caffeine, I footnote esoteric research conducted in the 1980s by NASA. Their scientists exposed spiders to different drugs and then observed the webs that they constructed.* Those drugs included LSD, speed (amphetamine), marijuana, and caffeine. The results, which speak for themselves, can be observed in figure 3. The researchers noted how strikingly incapable the spiders were in constructing anything resembling a normal or logical web that would be of any functional use when given caffeine, even relative to other potent drugs tested.

Figure 3: Effects of Various Drugs on Spider Web Building

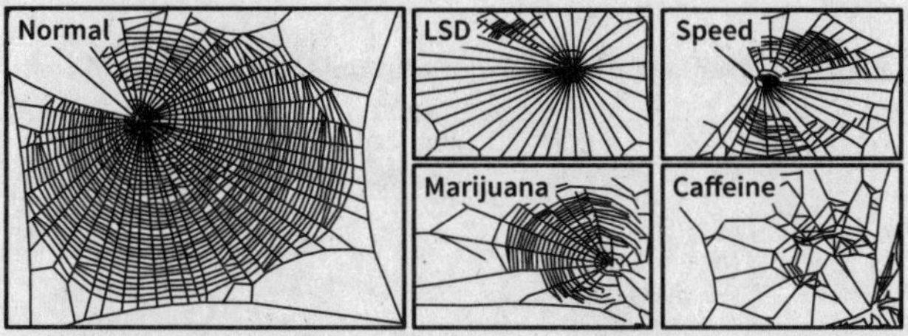

It is worth pointing out that caffeine is a stimulant drug. Caffeine is also the only addictive substance that we readily give to our children and teens—the consequences of which we will return to later in the book.

*R. Noever, J. Cronise, and R. A. Relwani, "Using spider-web patterns to determine toxicity," NASA Tech Briefs 19, no. 4 (1995): 82; and Peter N. Witt and Jerome S. Rovner, *Spider Communication: Mechanisms and Ecological Significance* (Princeton University Press, 1982).

IN STEP, OUT OF STEP

Setting caffeine aside for a moment, you may have assumed that the two governing forces that regulate your sleep—the twenty-four-hour circadian rhythm of the suprachiasmatic nucleus and the sleep-pressure signal of adenosine—communicate with each other so as to unite their influences. In actual fact, they don't. They are two distinct and separate systems that are ignorant of each other. They are not coupled; though, they are usually aligned.

Figure 4 encompasses forty-eight hours of time from left to right—two days and two nights. The dotted line in the figure is the circadian rhythm, also known as Process-C. Like a sine wave, it reliably and repeatedly rises and falls, and then rises and falls once more. Starting on the far left of the figure, the circadian rhythm begins to increase its activity a few hours before you wake up. It infuses the brain and body with an alerting energy signal. Think of it like a rousing marching band approaching from a distance. At first, the signal is faint, but gradually it builds, and builds, and builds with time. By early afternoon in most healthy adults, the activating signal from the circadian rhythm peaks.

Figure 4: The Two Factors Regulating Sleep and Wakefulness

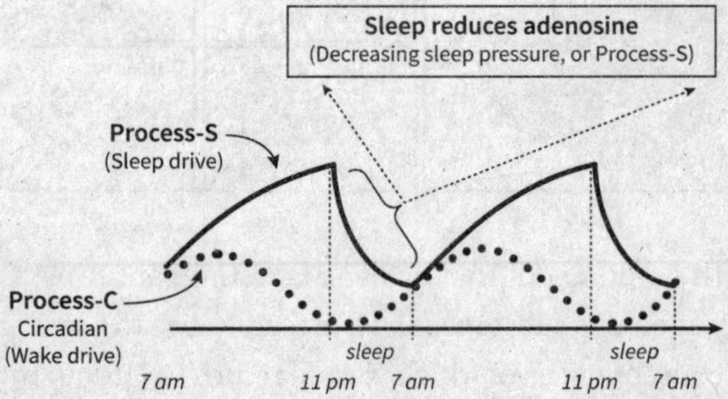

Now let us consider what is happening to the other sleep-controlling factor: adenosine. Adenosine creates a pressure to sleep, also known as Process-S. Represented by the solid line in figure 4, the longer you are awake, the more adenosine builds up, creating an increasing urge (pressure) to sleep. By mid- to late morning, you have only been awake for a handful of hours. As a result, adenosine concentrations have increased only a little. Furthermore, the circadian rhythm is on its powerful upswing of alertness. This combination of strong activating output from the circadian rhythm together with low levels of adenosine result in a delightful sensation of being wide awake. (Or at least it should, so long as your sleep was of good quality and sufficient length the night before. If you feel as though you could fall asleep easily midmorning, you are very likely not getting enough sleep, or the quality of your sleep is insufficient.) The distance between the curved lines above will be a direct reflection of your desire to sleep. The larger the distance between the two, the greater your sleep desire.

For example, at eleven a.m., after having woken up at eight a.m., there is only a small distance between the dotted line (circadian rhythm) and solid line (sleep pressure), illustrated by the vertical double arrow in figure 5. This minimal difference means there is a weak sleep drive, and a strong urge to be awake and alert.

Figure 5: The Urge to Be Awake

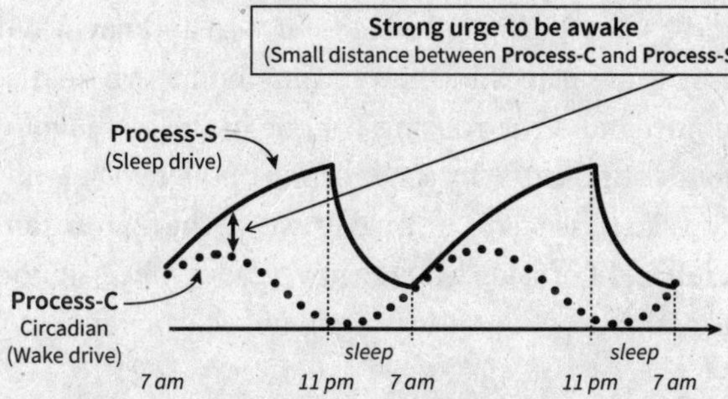

However, by eleven p.m. it's a very different situation, as illustrated in figure 6. You've now been awake for fifteen hours and your brain is drenched in high concentrations of adenosine (note how the solid line in the figure has risen sharply). In addition, the dotted line of the circadian rhythm is descending, powering down your activity and alertness levels. As a result, the difference between the two lines has grown large, reflected in the long vertical double arrow in figure 6. This powerful combination of abundant adenosine (high sleep pressure) and declining circadian rhythm (lowered activity levels) triggers a strong desire for sleep.

Figure 6: The Urge to Sleep

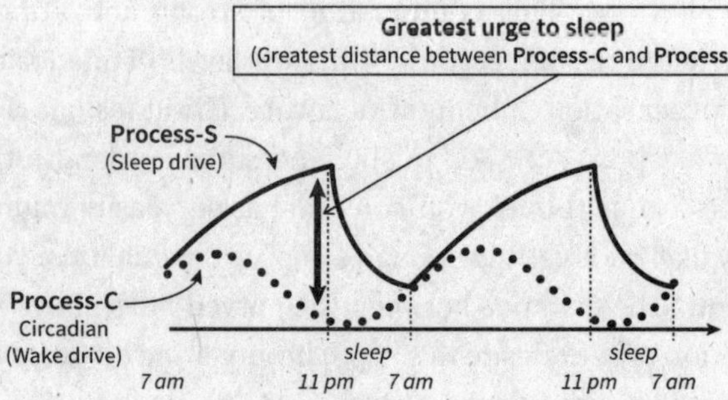

What happens to all of the accumulated adenosine once you do fall asleep? During sleep, a mass evacuation gets under way as the brain has the chance to degrade and remove the day's adenosine. Across the night, sleep lifts the heavy weight of sleep pressure, lightening the adenosine load. After approximately eight hours of healthy sleep in an adult, the adenosine purge is complete. Just as this process is ending, the marching band of your circadian activity rhythm has fortuitously returned, and its energizing influence starts to approach. When these two processes trade places in the morning hours, wherein adenosine has been removed and the rousing volume of the circadian rhythm is becoming louder (indicated by the meeting of the two lines in figure 6), we naturally wake up (seven a.m. on day two, in the figure example). Following that full night of sleep, you are now ready to face another sixteen hours of wakefulness with physical vigor and sharp brain function.

INDEPENDENCE DAY, AND NIGHT

Have you ever pulled an "all-nighter"—forgoing sleep and remaining awake throughout the following day? If you have, and can remember much of anything about it, you may recall that there were times when you felt truly miserable and sleepy, yet there were other moments when, despite having been awake for longer, you paradoxically felt *more* alert. Why? I don't advise anyone to conduct this self-experiment, but assessing a person's alertness across twenty-four hours of total sleep deprivation is one way that scientists can demonstrate that the two forces determining when you want to be awake and asleep—the twenty-four-hour circadian rhythm and the sleepiness signal of adenosine—are independent, and can be decoupled from their normal lockstep.

Let's consider figure 7, showing the same forty-eight-hour slice of time and the two factors in question: the twenty-four-hour circadian rhythm and the sleep pressure signal of adenosine, and how much distance there is between them. In this scenario, our volunteer is going to stay awake all night and all day. As the night of sleep deprivation marches forward, the sleep pressure of adenosine (upper line) rises progressively, like the rising water level in a plugged sink when a faucet has been left on. It will not decline across the night. It cannot, since sleep is absent.

Figure 7: The Ebb and Flow of Sleep Deprivation

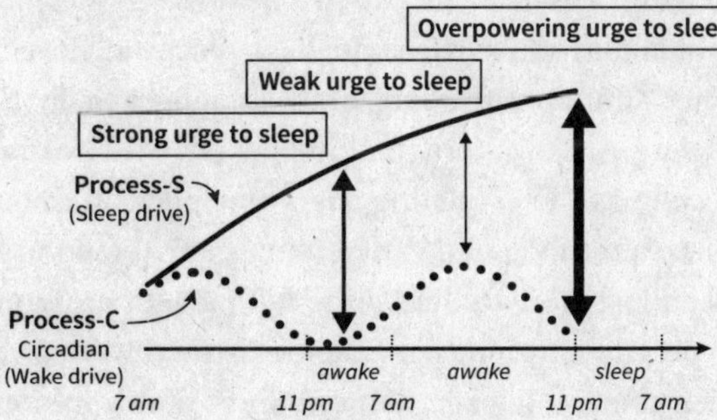

By remaining awake, and blocking access to the adenosine drain that sleep opens up, the brain is unable to rid itself of the chemical sleep pressure. The mounting adenosine levels continue to rise. This should mean that the longer you are awake, the sleepier you feel. But that's not true. Though you will feel increasingly sleepy throughout the nighttime phase, hitting a low point in your alertness around five to six a.m., thereafter, you'll catch a second wind. How is this possible when adenosine levels and corresponding sleep pressure continue to increase?

The answer resides with your twenty-four-hour circadian rhythm, which offers a brief period of salvation from sleepiness. Unlike sleep pressure, your circadian rhythm pays no attention to whether you are asleep or awake. Its slow, rhythmic countenance continues to fall and rise strictly on the basis of what time of night or day it is. No matter what state of adenosine sleepiness pressure exists within the brain, the twenty-four-hour circadian rhythm cycles on as per usual, oblivious to your ongoing lack of sleep.

If you look at figure 7 once again, the graveyard-shift misery you experience around six a.m. can be explained by the combination of high adenosine sleep pressure and your circadian rhythm reaching its lowest point. The vertical distance separating these two lines at three a.m. is large, indicated by the first vertical arrow in the figure. But if you can make it past this alertness low point, you're in for a rally. The morning rise of the circadian rhythm comes to your rescue, marshaling an alerting boost throughout the morning that temporarily offsets the rising levels of adenosine sleep pressure. As your circadian rhythm hits its peak around eleven a.m., the vertical distance between the two respective lines in figure 7 has been decreased.

The upshot is that you will feel much *less* sleepy at eleven a.m. than you did at three a.m., despite being awake for longer. Sadly, this second wind doesn't last. As the afternoon lumbers on, the circadian rhythm begins to decline as the escalating adenosine piles on the sleep pressure. Come late afternoon and early evening, any temporary alertness boost has been lost. You are hit by the full force of an immense adenosine sleep pressure. By nine p.m., there exists a towering vertical distance between the two lines in figure 7. Short of intravenous caffeine or amphetamine, sleep will have its way, wrestling your brain from the now weak grip of blurry wakefulness, blanketing you in slumber.

AM I GETTING ENOUGH SLEEP?

Setting aside the extreme case of sleep deprivation, how do you know whether you're routinely getting enough sleep? While a clinical sleep assessment is needed to thoroughly address this issue, an easy rule of thumb is to answer two simple questions. First, after waking up in the morning, could you fall back asleep at ten or eleven a.m.? If the answer is "yes," you are likely not getting sufficient sleep quantity and/or quality. Second, can you function optimally without caffeine before noon? If the answer is "no," then you are most likely self-medicating your state of chronic sleep deprivation.

Both of these signs you should take seriously and seek to address your sleep deficiency. They are topics, and a question, that we will cover in depth in chapters 13 and 14 when we speak about the factors that prevent and harm your sleep, as well as insomnia and effective treatments. In general, these un-refreshed feelings that compel a person to fall back asleep midmorning, or require the boosting of alertness with caffeine, are usually due to individuals not giving themselves adequate sleep opportunity time—at least eight or nine hours in bed. When you don't get enough sleep, one consequence among many is that adenosine concentrations remain too high. Like an outstanding debt on a loan, come the morning, some quantity of yesterday's adenosine remains. You then carry that outstanding sleepiness balance throughout the following day. Also like a loan in arrears, this sleep debt will continue to accumulate. You cannot hide from it. The debt will roll over into the next payment cycle, and the next, and the next, producing a condition of prolonged, chronic sleep deprivation from one day to another. This outstanding sleep obligation results in a feeling of chronic fatigue, manifesting in many forms of mental and physical ailments that are now rife throughout industrialized nations.

Other questions that can draw out signs of insufficient sleep are: If you didn't set an alarm clock, would you sleep past that time? (If so, you need more sleep than you are giving yourself.) Do you find yourself at your computer screen reading and then rereading (and perhaps rereading again) the same sentence? (This is often a sign of a fatigued, underslept brain.) Do you sometimes forget what color the last few traffic lights were while driving? (Simple distraction is often the cause, but a lack of sleep is another culprit.)

Of course, even if you are giving yourself plenty of time to get a full night of shut-eye, next-day fatigue and sleepiness can still occur because you are suffering from an undiagnosed sleep disorder, of which there are now more than a hundred. The most common is insomnia, followed by sleep-disordered breathing, or sleep apnea, which includes heavy snoring. Should you suspect your sleep or that of anyone else to be disordered, resulting in daytime fatigue, impairment, or distress, speak to your doctor immediately and seek a referral to a sleep specialist. Most important in this regard: you may not need to seek out sleeping pills as your first option. You will realize why I say this come chapter 14, but please feel free to skip right to the section on sleeping pills in that chapter if you are a current user, or considering using sleeping pills in the immediate future.

In the event it helps, I have provided a link to a questionnaire that has been developed by sleep researchers that will allow you to determine your degree of sleep fulfillment.* Called SATED, it is easy to complete, and contains only five simple questions.

*https://www.ncbi.nlm.nih.gov/pmc/articles/PMC3902880/bin/aasm.37.1.9s1.tif (source: D. J. Buysse, "Sleep Health: Can we define it? Does it matter?" *SLEEP* 37, no. 1 [2014]: 9–17).

CHAPTER 3

Defining and Generating Sleep

Time Dilation and What We Learned from a Baby in 1952

Perhaps you walked into your living room late one night while chatting with a friend. You saw a family member (let's call her Jessica) lying still on the couch, not making a peep, body recumbent and head lolling to one side. Immediately, you turned to your friend and said, "*Shhhhh, Jessica's sleeping.*" But how did you know? It took a split second of time, yet there was little doubt in your mind about Jessica's state. Why, instead, did you not think Jessica was in a coma, or worse, dead?

SELF-IDENTIFYING SLEEP

Your lightning-quick judgment of Jessica being asleep was likely correct. And perhaps you accidentally confirmed it by knocking something over and waking her up. Over time, we have all become incredibly good at recognizing a number of signals that suggest that another individual is asleep. So reliable are these signs that there now exists a set of observable features that scientists agree indicate the presence of sleep in humans and other species.

The Jessica vignette illustrates nearly all of these clues. First, sleeping organisms adopt a stereotypical position. In land animals, this is often horizontal, as was Jessica's position on the couch. Second, and related, sleeping organisms have lowered muscle tone. This is most evident in the relaxation of postural (antigravity) skeletal muscles—those that keep you upright, preventing you from collapsing to the floor. As these muscles ease their tension in light and then deep sleep, the body will slouch down. A sleeping organism will be draped over whatever supports it underneath, most evident in Jessica's listing head position. Third, sleeping individuals show no overt displays of communication or responsivity. Jessica showed no signs of orienting to you as you entered the room, as she would have when awake. The fourth defining feature of sleep is that it's easily reversible, differentiating it from coma, anesthesia, hibernation, and death. Recall that upon knocking the item over in the room, Jessica awoke. Fifth, as we established in the previous chapter, sleep adheres to a reliable timed pattern across twenty-four hours, instructed by the circadian rhythm coming from the brain's suprachiasmatic nucleus pacemaker. Humans are diurnal, so we have a preference for being awake throughout the day and sleeping at night.

Now let me ask you a rather different question: How do you, yourself, know that you have slept? You make this self-assessment even more frequently than that of sleep in others. Each morning, with luck, you return to the waking world knowing that you have been asleep.* So sensitive is this self-assessment of sleep that you can go a step further, gauging when you've had good- or bad-quality sleep. This is another way of measuring sleep—a first-person phenomenological assessment distinct from signs that you use to determine sleep in another.

Here, also, there are universal indicators that offer a convincing conclusion of sleep—two, in fact. First is the loss of external awareness—you stop perceiving the outside world. You are no longer conscious of all that surrounds you, at least not explicitly. In actual fact, your ears are still "hearing"; your eyes, though closed, are still capable of "seeing." This is similarly true for the other sensory organs of the nose (smell), the tongue (taste), and the skin (touch).

All these signals still flood into the center of your brain, but it is here, in the sensory convergence zone, where that journey ends while you sleep. The signals are blocked by a perceptual barricade set up in a structure called the thalamus (*THAL-uh-muhs*). A smooth, oval-shaped object just smaller than a lemon, the thalamus is the sensory gate of the brain. The thalamus decides which sensory signals are allowed through

*Some people with a certain type of insomnia are not able to accurately gauge whether they have been asleep or awake at night. As a consequence of this "sleep misperception," they underestimate how much slumber they have successfully obtained—a condition that we will return to later in the book.

its gate, and which are not. Should they gain privileged passage, they are sent up to the cortex at the top of your brain, where they are consciously perceived. By locking its gates shut at the onset of healthy sleep, the thalamus imposes a sensory blackout in the brain, preventing onward travel of those signals up to the cortex. As a result, you are no longer consciously aware of the information broadcasts being transmitted from your outer sense organs. At this moment, your brain has lost waking contact with the outside world that surrounds you. Said another way, you are now asleep.

The second feature that instructs your own, self-determined judgment of sleep is a sense of time distortion experienced in two contradictory ways. At the most obvious level, you lose your conscious sense of time when you sleep, tantamount to a chronometric void. Consider the last time you fell asleep on an airplane. When you woke up, you probably checked a clock to see how long you had been asleep. Why? Because your explicit tracking of time was ostensibly lost while you slept. It is this feeling of a time cavity that, in waking retrospect, makes you confident you've been asleep.

But while your *conscious* mapping of time is lost during sleep, at a *non-conscious* level, time continues to be cataloged by the brain with incredible precision. I'm sure you have had the experience of needing to wake up the next morning at a specific time. Perhaps you had to catch an early-morning flight. Before bed, you diligently set your alarm for 6:00 a.m. Miraculously, however, you woke up at 5:58 a.m., unassisted, right before the alarm. Your brain, it seems, is still capable of logging time with quite remarkable precision while asleep. Like so many other operations occurring within the brain, you simply don't have explicit access to this accurate time knowledge during sleep. It all flies below the radar of consciousness, surfacing only when needed.

One last temporal distortion deserves mention here—that of time dilation in dreams, beyond sleep itself. Time isn't quite time within dreams. It is most often elongated. Consider the last time you hit the snooze button on your alarm, having been woken from a dream. Mercifully, you are giving yourself another delicious five minutes of sleep. You go right back to dreaming. After the allotted five minutes, your alarm clock faithfully sounds again, yet that's not what it felt like to you. During those five minutes of actual time, you may have felt like you were dreaming for an hour, perhaps more. Unlike the phase of sleep where you are not dreaming, wherein you lose all awareness of time, in dreams, you continue to have a sense of time. It's simply not particularly accurate—more often than not dream time is stretched out and prolonged relative to real time.

Although the reasons for such time dilation are not fully understood, recent experimental recordings of brain cells in rats give tantalizing clues. In the experiment, rats were allowed to run around a maze. As the rats learned the spatial layout, the researchers recorded signature patterns of brain-cell firing. The scientists did not stop recording from these memory-imprinting cells when the rats subsequently fell asleep. They continued to eavesdrop on the brain during the different stages of slumber, including rapid eye movement (REM) sleep, the stage in which humans principally dream.

The first striking result was that the signature pattern of brain-cell firing that occurred as the rats were learning the maze subsequently reappeared during sleep, over and over again. That is, memories were being "replayed" at the level of brain-cell activity as the rats snoozed. The second, more striking finding was the speed of replay. During REM sleep, the memories were being replayed far more slowly: at just half or quarter the speed of that measured when the rats were awake and learning the maze. This slow neural recounting of the day's events is the best evidence we have to date explaining our own protracted experience of time in human REM sleep. This dramatic deceleration of neural time may be the reason we believe our dream life lasts far longer than our alarm clocks otherwise assert.

AN INFANT REVELATION—TWO TYPES OF SLEEP

Though we have all determined that someone is asleep, or that we have been asleep, the gold-standard scientific verification of sleep requires the recording of signals, using electrodes, arising from three different regions: (1) brainwave activity, (2) eye movement activity, and (3) muscle activity. Collectively, these signals are grouped together under the blanket term "polysomnography" (PSG), meaning a readout (*graph*) of sleep (*somnus*) that is made up of multiple signals (*poly*).

It was using this collection of measures that arguably the most important discovery in all of sleep research was made in 1952 at the University of Chicago by Eugene Aserinsky (then a graduate student) and Professor Nathaniel Kleitman, famed for the Mammoth Cave experiment discussed in chapter 2.

Aserinsky had been carefully documenting the eye movement patterns of human infants during the day and night. He noticed that there were periods of sleep when the eyes would rapidly dart from side to side underneath their lids. Furthermore, these sleep phases were always accompanied by remarkably active brainwaves, almost identical to those observed from a brain that is wide awake. Sandwiching these earnest phases of active sleep were longer swaths of time when the eyes would calm and rest still. During these quiescent time periods, the brainwaves would also become calm, slowly ticking up and down.

As if that weren't strange enough, Aserinsky also observed that these two phases of slumber (sleep with eye movements, sleep with no eye movements) would repeat in a somewhat regular pattern throughout the night, over, and over, and over again.

With classic professorial skepticism, his mentor, Kleitman, wanted to see the results replicated before he would entertain their validity. With his propensity for including his nearest and dearest in his experimentation, he chose his infant daughter, Ester, for this investigation. The findings held up. At that moment Kleitman and Aserinsky realized the profound discovery they had made: humans don't just sleep, but cycle through two completely different types of sleep. They named these sleep stages based on their defining ocular features: non–rapid eye movement, or NREM, sleep, and rapid eye movement, or REM, sleep.

Together with the assistance of another graduate student of Kleitman's at the time, William Dement, Kleitman and Aserinsky further demonstrated that REM sleep, in which brain activity was almost identical to that when we are awake, was intimately connected to the experience we call dreaming, and is often described as dream sleep.

NREM sleep received further dissection in the years thereafter, being subdivided into four separate stages, unimaginatively named NREM stages 1 to 4 (we sleep researchers are a creative bunch), increasing in their depth. Stages 3 and 4 are therefore the deepest stages of NREM sleep you experience, with "depth" being defined as the increasing difficulty required to wake an individual out of NREM stages 3 and 4, compared with NREM stages 1 or 2.

THE SLEEP CYCLE

In the years since Ester's slumber revelation, we have learned that the two stages of sleep—NREM and REM—play out in a recurring, push-pull battle for brain domination across the night. The cerebral war between the two is won and lost every ninety minutes,* ruled first by NREM sleep, followed by the comeback of REM sleep. No sooner has the battle finished than it starts anew, replaying every ninety minutes. Tracing this remarkable roller-coaster ebb and flow across the night reveals the quite beautiful cycling architecture of sleep, depicted in figure 8.

On the vertical axis are the different brain states, with Wake at the top, then REM sleep, and then the descending stages of NREM sleep, stages 1 to 4. On the horizontal axis is time of night, starting on the left at about eleven p.m. through until seven a.m. on the right. The technical name for this graphic is a hypnogram (a sleep graph).

Figure 8: The Architecture of Sleep

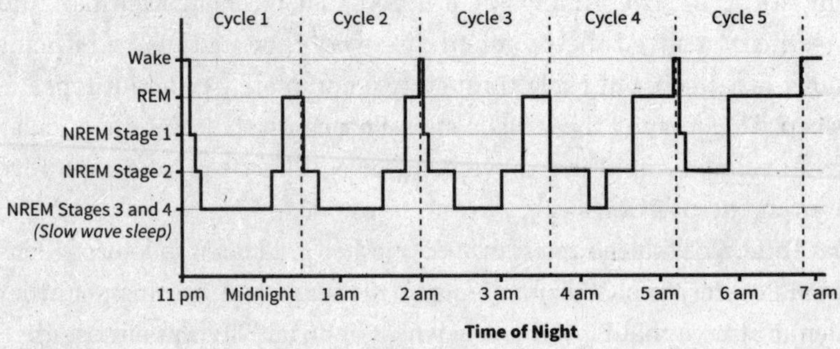

*Different species have different NREM–REM cycle lengths. Most are shorter than that of humans. The functional purpose of the cycle length is another mystery of sleep. To date, the best predictor of NREM–REM sleep cycle length is the width of the brain stem, with those species possessing wider brain stems having longer cycle lengths.

Had I not added the vertical dashed lines demarcating each ninety-minute cycle, you may have protested that you could not see a regularly repeating ninety-minute pattern. At least not the one you were expecting from my description above. The cause is another peculiar feature of sleep: a lopsided profile of sleep stages. While it is true that we flip-flop back and forth between NREM and REM sleep throughout the night every ninety minutes, the ratio of NREM sleep to REM sleep within each ninety-minute cycle changes dramatically across the night. In the first half of the night, the vast majority of our ninety-minute cycles are consumed by deep NREM sleep, and very little REM sleep, as can be seen in cycle 1 of the figure above. But as we transition through into the second half of the night, this seesaw balance shifts, with most of the time dominated by REM sleep, with little, if any, deep NREM sleep. Cycle 5 is a perfect example of this REM-rich type of sleep.

Why did Mother Nature design this strange, complex equation of unfolding sleep stages? Why cycle between NREM and REM sleep over and over? Why not obtain all of the required NREM sleep first, followed by all of the necessary REM sleep second? Or vice versa? If that's too much a gamble on the off chance that an animal only obtains a partial night of sleep at some point, then why not keep the ratio within each cycle the same, placing similar proportions of eggs in both baskets, as it were, rather than putting most of them in one early on, and then inverting that imbalance later in the night? Why vary it? It sounds like an exhausting amount of evolutionary hard work to have designed such a convoluted system, and put it into biological action.

We have no scientific consensus as to why our sleep (and that of all other mammals and birds) cycles in this repeatable but dramatically asymmetric pattern, though a number of theories exist. One theory I have offered is that the uneven back-and-forth interplay between NREM and REM sleep is necessary to elegantly remodel and update our neural circuits at night, and in doing so manage the finite storage space within the brain. Forced by the known storage capacity imposed by a set number of neurons and connections within their memory structures, our brains must find the "sweet spot" between retention of old information and leaving sufficient room for the new. Balancing this storage equation requires identifying which memories are fresh and salient, and which memories that currently exist are overlapping, redundant, or simply no longer relevant.

As we will discover in chapter 6, a key function of deep NREM sleep, which predominates early in the night, is to do the work of weeding out and removing unnecessary neural connections. In contrast, the dreaming stage of REM sleep, which prevails later in the night, plays a role in strengthening those connections.

Combine these two, and we have at least one parsimonious explanation for why the two types of sleep cycle across the night, and why those cycles are initially dominated by NREM sleep early on, with REM sleep reigning supreme in the second half of the night. Consider the creation of a piece of sculpture from a block of clay. It starts with placing a large amount of raw material onto a pedestal (that entire mass of stored autobiographical memories, new and old, offered up to sleep each night). Next comes an initial and extensive removal of superfluous matter (long stretches of NREM sleep), after which brief intensification of early details can be made (short REM periods). Following this first session, the culling hands return for a second round of deep excavation (another long NREM-sleep phase), followed by a little more enhancing of some fine-grained structures that have emerged (slightly more REM sleep). After several more cycles of work, the balance of sculptural need has shifted. All core features have been hewn from the original mass of raw material. With only the important clay remaining, the work of the sculptor, and the tools required, must shift toward the goal of strengthening the elements and enhancing features of that which remains (a dominant need for the skills of REM sleep, and little work remaining for NREM sleep).

In this way, sleep may elegantly manage and solve our memory storage crisis, with the general excavatory force of NREM sleep dominating early, after which the etching hand of REM sleep blends, interconnects, and adds details. Since life's experience is ever changing, demanding that our memory catalog be updated ad infinitum, our autobiographical sculpture of stored experience is never complete. As a result, the brain always requires a new bout of sleep and its varied stages each night so as to auto-update our memory networks based on the events of the prior day. This account is one reason (of many, I suspect) explaining the cycling nature of NREM and REM sleep, and the imbalance of their distribution across the night.

A danger resides in this sleep profile wherein NREM dominates early in the night, followed by an REM sleep dominance later in the morning, one of which most of the general public are unaware. Let's say that you go to bed this evening at midnight. But instead of waking up at eight a.m., getting a full eight hours of sleep, you must wake up at six a.m. because of an early-morning meeting or because you are an athlete whose coach demands early-morning practices. What percent of sleep will you lose? The logical answer is 25 percent, since waking up at six a.m. will lop off two hours of sleep from what would otherwise be a normal eight hours. But that's not entirely true. Since your brain desires most of its REM sleep in the last part of the night, which is to say the late-morning hours, you may lose 60 to 90 percent of all your REM sleep, even though you are losing 25 percent of your total sleep time. It works both ways. If you wake up at ten a.m., but don't go to bed until four a.m., then you will lose a significant amount of your normal deep NREM sleep. Similar to an unbalanced diet in which you only eat carbohydrates and are left malnourished by the absence of protein, short-changing the brain of either NREM or REM sleep—both of which serve critical, though different, brain and body functions—results in a myriad of physical and mental ill health, as we will see in later chapters. When it comes to sleep, there is no such thing as burning the candle at both ends—or even at one end.

HOW YOUR BRAIN GENERATES SLEEP

If I brought you into my sleep laboratory this evening at the University of California, Berkeley, placed electrodes on your head and face, and let you fall asleep, what would your sleeping brainwaves look like? How different would those patterns of brain activity be to those you are experiencing right now, as you read this sentence, awake? How do these different electrical brain changes explain why you are conscious in one state (wake), non-conscious in another (NREM sleep), and delusionally conscious, or dreaming, in the third (REM sleep)?

Figure 9: The Brainwaves of Wake and Sleep

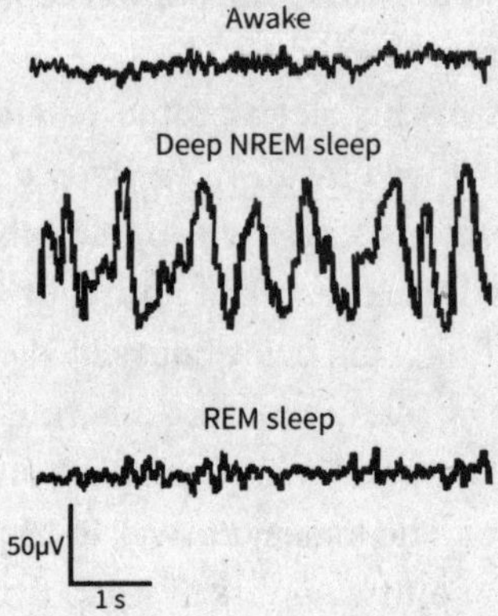

Assuming you are a healthy young/midlife adult (we will discuss sleep in childhood, old age, and disease a little later), the three wavy lines in figure 9 reflect the different types of electrical activity I would record from your brain. Each line represents thirty seconds of brainwave activity from these three different states: (1) wakefulness, (2) deep NREM sleep, and (3) REM sleep.

Prior to bed, your waking brain activity is frenetic, meaning that the brainwaves are cycling (going up and down) perhaps thirty or forty times per second, similar to a very fast drumbeat. This is termed "fast frequency" brain activity. Moreover, there is no reliable pattern to these brainwaves—that is, the drumbeat is not only fast, but also erratic. If I asked you to predict the next few seconds of the activity by tapping along to the beat, based on what came before, you would not be able to do so. The brainwaves are really that asynchronous—their drumbeat has no discernible rhythm. Even if I converted the brainwaves into sound (which I have done in my laboratory in a sonification-of-sleep project, and is eerie to behold), you would find it impossible to dance to. These are the electrical hallmarks of full wakefulness: fast-frequency, chaotic brainwave activity.

You may have been expecting your general brainwave activity to look beautifully coherent and highly synchronous while awake, matching the ordered pattern of your (mostly) logical thought during waking consciousness. The contradictory electrical chaos is explained by the fact that different parts of your waking brain are processing different pieces of information at different moments in time and in different ways. When summed together, they produce what appears to be a discombobulated pattern of activity recorded by the electrodes placed on your head.

As an analogy, consider a large football stadium filled with thousands of fans. Dangling over the middle of the stadium is a microphone. The individual people in the stadium represent individual brain cells, seated in different parts of the stadium, as they are clustered in different regions of the brain. The microphone is the electrode, sitting on top of the head—a recording device.

Before the game starts, all of the individuals in the stadium are speaking about different things at different times. They are not having the same conversation in sync. Instead, they are desynchronized in their individual discussions. As a result, the summed chatter that we pick up from the overhead microphone is chaotic, lacking a clear, unified voice.

When an electrode is placed on a subject's head, as done in my laboratory, it is measuring the summed activity of all the neurons below the surface of the scalp as they process different streams of information (sounds, sights, smells, feelings, emotions) at different moments in time and in different underlying locations. Processing that much information of such varied kinds means that your brainwaves are very fast, frenetic, and chaotic.

Once settled into bed at my sleep laboratory, with lights out and perhaps a few tosses and turns here and there, you will successfully cast off from the shores of wakefulness into sleep. First, you will wade out into the shallows of light NREM sleep: stages 1 and 2. Thereafter, you will enter the deeper waters of stages 3 and 4 of NREM sleep, which are grouped together under the blanket term "slow-wave sleep." Returning to the brainwave patterns of figure 9, and focusing on the middle line, you can understand why. In deep, slow-wave sleep, the up-and-down tempo of your brainwave activity dramatically decelerates, perhaps just two to four waves per second: ten times slower than the fervent speed of brain activity you were expressing while awake.

As remarkable, the slow waves of NREM are also far more synchronous and reliable than those of your waking brain activity. So reliable, in fact, that you could predict the next few bars of NREM sleep's electrical song based on those that came before. Were I to convert the deep rhythmic activity of your NREM sleep into sound and play it back to you in the morning (which we have also done for people in the same sonification-of-sleep project), you'd be able to find its rhythm and move in time, gently swaying to the slow, pulsing measure.

But something else would become apparent as you listened and swayed to the throb of deep-sleep brainwaves. Every now and then a new sound would be overlaid on top of the slow-wave rhythm. It would be brief, lasting only a few seconds, but it would always occur on the downbeat of the slow-wave cycle. You would perceive it as a quick trill of sound, not dissimilar to the strong rolling *r* in certain languages, such as Hindi or Spanish, or a very fast *purrr* from a pleased cat.

What you are hearing is a sleep spindle—a punchy burst of brainwave activity that often festoons the tail end of each individual slow wave. Sleep spindles occur during both the deep and the lighter stages of NREM sleep, even before the slow, powerful brainwaves of deep sleep start to rise up and dominate. One of their many functions is to operate like nocturnal soldiers who protect sleep by shielding the brain from external noises. The more powerful and frequent an individual's sleep spindles, the more resilient they are to external noises that would otherwise awaken the sleeper.

Returning to the slow waves of deep sleep, we have also discovered something fascinating about their site of origin, and how they sweep across the surface of the brain. Place your finger between your eyes, just above the bridge of your nose. Now slide it up your forehead about two inches. When you go to bed tonight, this is where most of your deep-sleep brainwaves will be generated: right in the middle of your frontal lobes. It is the epicenter, or hot spot, from which most of your deep, slow-wave sleep emerges. However, the waves of deep sleep do not radiate out in perfect circles. Instead, almost all of your deep-sleep brainwaves will travel in one direction: from the front of your brain to the back. They are like the sound waves emitted from a speaker, which predominantly travel in one direction, from the speaker outward (it is always louder in front of a speaker than behind it). And like a speaker broadcasting across a vast expanse, the slow waves that you generate tonight will gradually dissipate in strength as they make their journey to the back of the brain, without rebound or return.

Back in the 1950s and 1960s, as scientists began measuring these slow brainwaves, an understandable assumption was made: this leisurely, even lazy-looking electrical pace of brainwave activity must reflect a brain that is idle, or even dormant. It was a reasonable hunch considering that the deepest, slowest brainwaves of NREM sleep can resemble those we see in patients under anesthesia, or even those in certain forms of coma. But this assumption was utterly wrong. Nothing could be further from the truth. What you are actually experiencing during deep NREM sleep is one of the most epic displays of neural collaboration that we know of. Through an astonishing act of self-organization, many thousands of brain cells have all decided to unite and "sing," or fire, in time. Every time I watch this stunning act of neural synchrony occurring at night in my own research laboratory, I am humbled: sleep is truly an object of awe.

Returning to the analogy of the microphone dangling above the football stadium, consider the game of sleep now in play. The crowd—those thousands of brain cells—has shifted from their individual chitter-chatter before the game (wakefulness) to a unified state (deep sleep). Their voices have joined in a lockstep, mantra-like chant—the chant of deep NREM sleep. All at once they exuberantly shout out, creating the tall spike of brainwave activity, and then fall silent for several seconds, producing the deep, protracted trough of the wave. From our stadium microphone we pick up a clearly defined roar from the underlying crowd, followed by a long breath-pause. Realizing that the rhythmic *incantare* of deep NREM slow-wave sleep was actually a highly active, meticulously coordinated state of cerebral unity, scientists were forced to abandon any cursory notions of deep sleep as a state of semi-hibernation or dull stupor.

Understanding this stunning electrical harmony, which ripples across the surface of your brain hundreds of times each night, also helps explain your loss of external consciousness. It starts below the surface of the brain, within the thalamus. Recall that as we fall asleep, the thalamus—the sensory gate, seated deep in the middle of the brain—blocks the transfer of perceptual signals (sound, sight, touch, etc.) up to the top of the brain, or the cortex. By severing perceptual ties with the outside world, not only do we lose our sense of consciousness (explaining why we do not dream in deep NREM sleep, nor do we keep explicit track of time), this also allows the cortex to "relax" into its default mode of functioning. That default mode is what we call deep slow-wave sleep. It is an active, deliberate, but highly synchronous state of brain activity. It is a near state of nocturnal cerebral meditation, though I should note that it is very different from the brainwave activity of waking meditative states.

In this shamanistic state of deep NREM sleep can be found a veritable treasure trove of mental and physical benefits for your brain and body, respectively—a bounty that we will fully explore in chapter 6. However, one brain benefit—the saving of memories—deserves further mention at this moment in our story, as it serves as an elegant example of what those deep, slow brainwaves are capable of.

Have you ever taken a long road trip in your car and noticed that at some point in the journey, the FM radio stations you've been listening to begin dropping out in signal strength? In contrast, AM radio stations remain solid. Perhaps you've driven to a remote location and tried and failed to find a new FM radio station. Switch over to the AM band, however, and several broadcasting channels are still available. The explanation lies in the radio waves themselves, including the two different speeds of the FM and AM transmissions. FM uses faster-frequency radio waves that go up and down many more times per second than AM radio waves. One advantage of FM radio waves is that they can carry higher, richer loads of information, and hence they sound better. But there's a big disadvantage: FM waves run out of steam quickly, like a muscle-bound sprinter who can only cover short distances. AM broadcasts employ a much slower (longer) radio wave, akin to a lean long-distance runner. While AM radio waves cannot match the muscular, dynamic quality of FM radio, the pedestrian pace of AM radio waves gives them the ability to cover vast distances with less fade. Longer-range broadcasts are therefore possible with the slow waves of AM radio, allowing far-reaching communication between very distant geographic locations.

As your brain shifts from the fast-frequency activity of waking to the slower, more measured pattern of deep NREM sleep, the very same long-range communication advantage becomes possible. The steady, slow, synchronous waves that sweep across the brain during deep sleep open up communication possibilities between distant regions of the brain, allowing them to collaboratively send and receive their different repositories of stored experience.

In this regard, you can think of each individual slow wave of NREM sleep as a courier, able to carry packets of information between different anatomical brain centers. One benefit of these traveling deep-sleep brainwaves is a file-transfer process. Each night, the long-range brainwaves of deep sleep will move memory packets (recent experiences) from a short-term storage site, which is fragile, to a more permanent, and thus safer, long-term storage location. We therefore consider waking brainwave activity as that principally concerned with the *reception* of the outside sensory world, while the state of deep NREM slow-wave sleep donates a state of inward *reflection*—one that fosters information transfer and the distillation of memories.

If wakefulness is dominated by reception, and NREM sleep by reflection, what, then, happens during REM sleep—the dreaming state? Returning to figure 9, the last line of electrical brainwave activity is that which I would observe coming from your brain in the sleep lab as you entered into REM sleep. Despite being asleep, the associated brainwave activity bears no resemblance to that of deep NREM slow-wave sleep (the middle line in the figure). Instead, REM sleep brain activity is an almost perfect replica of that seen during attentive, alert wakefulness—the top line in the figure. Indeed, recent MRI scanning studies have found that there are individual parts of the brain that are up to 30 percent more active during REM sleep than when we are awake!

For these reasons, REM sleep has also been called paradoxical sleep: a brain that appears awake, yet a body that is clearly asleep. It is often impossible to distinguish REM sleep from wakefulness using just electrical brainwave activity. In REM sleep, there is a return of the same faster-frequency brainwaves that are once again desynchronized. The many thousands of brain cells in your cortex that had previously unified in a slow, synchronized chat during deep NREM sleep have returned to frantically processing different informational pieces at different speeds and times in different brain regions—typical of wakefulness. But you're not awake. Rather, you are sound asleep. So what information is being processed, since it is certainly not information from the outside world at that time?

As is the case when you are awake, the sensory gate of the thalamus once again swings open during REM sleep. But the nature of the gate is different. It is not sensations from the outside that are allowed to journey to the cortex during REM sleep. Rather, signals of emotions, motivations, and memories (past and present) are all played out on the big screens of our visual, auditory, and kinesthetic sensory cortices in the brain. Each and every night, REM sleep ushers you into a preposterous theater wherein you are treated to a bizarre, highly associative carnival of autobiographical themes. When it comes to information processing, think of the wake state principally as *reception* (experiencing and constantly learning the world around you), NREM sleep as *reflection* (storing and strengthening those raw ingredients of new facts and skills), and REM sleep as *integration* (interconnecting these raw ingredients with each other, with all past experiences, and, in doing so, building an ever more accurate model of how the world works, including innovative insights and problem-solving abilities).

Since the electrical brainwaves of REM sleep and wake are so similar, how can I tell which of the two you are experiencing as you lie in the bedroom of the sleep laboratory next to the control room? The telltale player in this regard is your body—specifically its muscles.

Before putting you to bed in the sleep laboratory, we would have applied electrodes to your body, in addition to those we affix to your head. While awake, even lying in bed and relaxed, there remains a degree of overall tension, or tone, in your muscles. This steady muscular hum is easily detected by the electrodes listening in on your body. As you pass into NREM sleep, some of that muscle tension disappears, but much remains. Gearing up for the leap into REM sleep, however, an impressive change occurs. Mere seconds before the dreaming phase begins, and for as long as that REM-sleep period lasts, you are completely paralyzed. There is no tone in the voluntary muscles of your body. None whatsoever. If I were to quietly come into the room and gently lift up your body without waking you, it would be completely limp, like a rag doll. Rest assured that your *involuntary* muscles—those that control automatic operations such as breathing—continue to operate and maintain life during sleep. But all other muscles become lax.

This feature, termed "atonia" (an absence of tone, referring here to the muscles), is instigated by a powerful disabling signal that is transmitted down the full length of your spinal cord from your brain stem. Once put in place, the postural body muscles, such as the biceps of your arms and the quadriceps of your legs, lose all tension and strength. No longer will they respond to commands from your brain. You have, in effect, become an embodied prisoner, incarcerated by REM sleep. Fortunately, after serving the detention sentence of the REM-sleep cycle, your body is freed from physical captivity as the REM-sleep phase ends. This striking dissociation during the dreaming state, where the brain is highly active but the body is immobilized, allows sleep scientists to easily recognize—and therefore separate—REM-sleep brainwaves from wakeful ones.

Why did evolution decide to outlaw muscle activity during REM sleep? Because by eliminating muscle activity you are prevented from acting out your dream experience. During REM sleep, there is a nonstop barrage of motor commands swirling around the brain, and they underlie the movement-rich experience of dreams. Wise, then, of Mother Nature to have tailored a physiological straitjacket that forbids these fictional movements from becoming reality, especially considering that you've stopped consciously perceiving your surroundings. You can well imagine the calamitous upshot of falsely enacting a dream fight, or a frantic sprint from an approaching dream foe, while your eyes are closed and you have no comprehension of the world around you. It wouldn't take long before you quickly left the gene pool. The brain paralyzes the body so the mind can dream safely.

How do we know these movement commands are actually occurring while someone dreams, beyond the individual simply waking up and telling you they were having a running dream or a fighting dream? The sad answer is that this paralysis mechanism can fail in some people, particularly later in life. Consequentially, they convert these dream-related motor impulses into real-world physical actions. As we shall read about in chapter 11, the repercussions can be tragic.

Finally, and not to be left out of the descriptive REM-sleep picture, is the very reason for its name: corresponding rapid eye movements. Your eyes remain still in their sockets during deep NREM sleep.* Yet electrodes that we place above and below your eyes tell a very different ocular story when you begin to dream: the very same story that Kleitman and Aserinsky unearthed in 1952 when observing infant sleep. During REM sleep, there are phases when your eyeballs will jag, with urgency, left-to-right, left-to-right, and so on. At first, scientists assumed that these rat-a-tat-tat eye movements corresponded to the tracking of visual experience in dreams. This is not true. Instead, the eye movements are intimately linked with the physiological creation of REM sleep, and reflect something even more extraordinary than the passive apprehension of moving objects within dream space. This phenomenon is chronicled in detail in chapter 9.

Are we the only creatures that experience these varied stages of sleep? Do any other animals have REM sleep? Do they dream? Let us find out.

*Oddly, during the transition from being awake into light stage 1 NREM sleep, the eyes will gently and very, very slowly start to roll in their sockets in synchrony, like two ocular ballerinas pirouetting in perfect time with each other. It is a hallmark indication that the onset of sleep is inevitable. If you have a bed partner, try observing their eyelids the next time they are drifting off to sleep. You will see the closed lids of the eyes deforming as the eyeballs roll around underneath. Parenthetically, should you choose to complete this suggested observational experiment, be aware of the potential ramifications. There is perhaps little else more disquieting than aborting one's transition into sleep, opening your eyes, and finding your partner's face looming over yours, gaze affixed.

CHAPTER 4

Ape Beds, Dinosaurs, and Napping with Half a Brain

Who Sleeps, How Do We Sleep, and How Much?

WHO SLEEPS

When did life start sleeping? Perhaps sleep emerged with the great apes? Maybe earlier, in reptiles or their aquatic antecedents, fish? Short of a time capsule, the best way to answer this question comes from studying sleep across different phyla of the animal kingdom, from the prehistoric to the evolutionarily recent. Investigations of this kind provide a powerful ability to peer far back in the historical record and estimate the moment when sleep first graced the planet. As the geneticist Theodosius Dobzhansky once said, "Nothing in biology makes sense except in light of evolution." For sleep, the illuminating answer turned out to be far earlier than anyone anticipated, and far more profound in ramification.

Every animal species carefully studied to date sleeps, or engages in something remarkably like it. This includes insects, such as flies, bees, cockroaches, and scorpions;* fish, from small perch to the largest sharks;† amphibians, such as frogs; and reptiles, such as turtles, Komodo dragons, and chameleons. Ascend the evolutionary ladder further and we find that all types of birds and mammals sleep: from shrews to parrots, kangaroos, polar bears, bats, and, of course, we humans.

---

*Proof of sleep in very small species, such as insects, in which recordings of electrical activity from the brain are impossible, is confirmed using the same set of behavioral features described in chapter 3, illustrated by the example of Jessica: immobility, reduced responsiveness to the outside world, easily reversible. A further criterion is that depriving the organism of what looks like sleep should result in an increased drive for more of it when you stop the annoying deprivation assault, reflecting "sleep rebound."

†It was once thought that sharks did not sleep, in part because they never closed their eyes. Indeed, they do have clear active and passive phases that resemble wake and sleep. We now know that the reason they never close their eyes is because they have no eyelids.

Even invertebrates, such as primordial mollusks and echinoderms, and even very primitive worms, enjoy periods of what appears to be slumber. In these phases, affectionately termed "lethargus," they, like humans, become unresponsive to external stimuli. And just as we fall asleep faster and sleep more soundly when sleep-deprived, so, too, do worms, defined by their degree of insensitivity to prods from experimenters.

How "old" does this make sleep? Worms emerged during the Cambrian explosion: at least 500 million years ago. That is, worms (and sleep by association) predate all vertebrate life. This includes dinosaurs, which, by inference, are likely to have slept. Imagine diplodocuses and triceratopses all comfortably settling in for a night of full repose!

Regress evolutionary time still further and we have discovered that even some of the very simplest forms of unicellular organisms, such as bacteria, have active and passive phases that correspond to the light-dark cycle of our planet. It is a pattern that we now believe to be the precursor of our own circadian rhythm, and with it, wake and sleep.

Many of the explanations for why we sleep circle around a common, and perhaps erroneous, idea: sleep is the state we must enter in order to fix that which has been upset by wake. But what if we turned this argument on its head? What if sleep is so useful—so physiologically beneficial to every aspect of our being—that the real question is: Why did life ever bother to wake up? Considering how biologically damaging the state of wakefulness can often be, that is the true evolutionary puzzle here, not sleep. Adopt this perspective, and we can pose a very different theory: sleep was the first state of life on this planet, and it was from sleep that wakefulness emerged. It may be a preposterous hypothesis, but personally I do not think it to be entirely unreasonable.

Whichever of these two theories is true, what we know for certain is that sleep is of ancient origin. It appeared with the very earliest forms of planetary life. Like other rudimentary features, such as DNA, sleep may be a common bond uniting every creature in the animal kingdom. A long-lasting commonality, yes; however, there are truly remarkable differences in sleep from one species to another. Four such differences, in fact.

ONE OF THESE THINGS IS NOT LIKE THE OTHER

Elephants need half as much sleep as humans, requiring just four hours of slumber each day. Tigers and lions devour fifteen hours of daily sleep. The brown bat outperforms all other mammals, being awake for just five hours each day while sleeping nineteen hours. *Total amount of time* is one of the most conspicuous differences in how organisms sleep.

You'd imagine the reason for such clear-cut variation in sleep need is obvious. It isn't. None of the likely contenders—body size, prey/predator status, diurnal/nocturnal—usefully explains the difference in sleep need across species. Surely sleep time is at least similar within any one phylogenetic category, since they share much of their genetic code. It is certainly true for other basic traits within phyla, such as sensory capabilities, methods of reproduction, and even degree of intelligence. Yet sleep violates this reliable pattern. Squirrels and degus are part of the same family group (rodents), yet they could not be more dissimilar in sleep need. The former sleeps twice as long as the latter—15.9 hours for the squirrel versus 7.7 hours for the degu. Conversely, you can find near-identical sleep times in utterly different family groups. The humble guinea pig and the precocious baboon, for example, which are of markedly different phylogenetic orders, not to mention physical sizes, sleep precisely the same amount: 9.4 hours.

So what does explain the difference in sleep time (and perhaps need) from species to species, or even within a genetically similar order? We're not entirely sure. The relationship between the size of the nervous system, the complexity of the nervous system, and total body mass appears to be a somewhat meaningful predictor, with increasing brain complexity *relative* to body size resulting in greater sleep amounts. While weak and not entirely consistent, this relationship suggests that one evolutionary function that demands more sleep is the need to service an increasingly complex nervous system. As millennia unfolded and evolution crowned its (current) accomplishment with the genesis of the brain, the demand for sleep only increased, tending to the needs of this most precious of all physiological apparatus.

Yet this is not the whole story—not by a good measure. Numerous species deviate wildly from the predictions made by this rule. For example, an opossum, which weighs almost the same as a rat, sleeps 50 percent longer, clocking an average of eighteen hours each day. The opossum is just one hour shy of the animal kingdom record for sleep amount currently held by the brown bat, who, as previously mentioned, racks up a whopping nineteen hours of sleep each day.

There was a moment in research history when scientists wondered if the measure of choice—total minutes of sleep—was the wrong way of looking at the question of why sleep varies so considerably across species. Instead, they suspected that assessing sleep *quality*, rather than *quantity* (time), would shed some light on the mystery. That is, species with superior quality of sleep should be able to accomplish all they need in a shorter time, and vice versa. It was a great idea, with the exception that, if anything, we've discovered the opposite relationship: those that sleep more have deeper, "higher"-quality sleep. In truth, the way quality is commonly assessed in these investigations (degree of unresponsiveness to the outside world and the continuity of sleep) is probably a poor index of the real biological measure of sleep quality: one that we cannot yet obtain in all these species. When we can, our understanding of the relationship between sleep quantity and quality across the animal kingdom will likely explain what currently appears to be an incomprehensible map of sleep-time differences.

For now, our most accurate estimate of why different species need different sleep amounts involves a complex hybrid of factors, such as dietary type (omnivore, herbivore, carnivore), predator/prey balance within a habitat, the presence and nature of a social network, metabolic rate, and nervous system complexity. To me, this speaks to the fact that sleep has likely been shaped by numerous forces along the evolutionary path, and involves a delicate balancing act between meeting the demands of waking survival (e.g., hunting prey/obtaining food in as short a time as possible, minimizing energy expenditure and threat risk), serving the restorative physiological needs of an organism (e.g., a higher metabolic rate requires greater "cleanup" efforts during sleep), and tending to the more general requirements of the organism's community.

Nevertheless, even our most sophisticated predictive equations remain unable to explain far-flung outliers in the map of slumber: species that sleep much (e.g., bats) and those that sleep little (e.g., giraffes, which sleep for just four to five hours). Far from being a nuisance, I feel these anomalous species may hold some of the keys to unlocking the puzzle of sleep need. They remain a delightfully frustrating opportunity for those of us trying to crack the code of sleep across the animal kingdom, and within that code, perhaps as yet undiscovered benefits of sleep we never thought possible.

TO DREAM OR NOT TO DREAM

Another remarkable difference in sleep across species is *composition.* Not all species experience all stages of sleep. Every species in which we can measure sleep stages experiences NREM sleep—the non-dreaming stage. However, insects, amphibians, fish, and most reptiles show no clear signs of REM sleep—the type associated with dreaming in humans. Only birds and mammals, which appeared later in the evolutionary timeline of the animal kingdom, have full-blown REM sleep. It suggests that dream (REM) sleep is the new kid on the evolutionary block. REM sleep seems to have emerged to support functions that NREM sleep alone could not accomplish, or that REM sleep was more efficient at accomplishing.

Yet as with so many things in sleep, there is another anomaly. I said that all mammals have REM sleep, but debate surrounds cetaceans, or aquatic mammals. Certain of these ocean-faring species, such as dolphins and killer whales, buck the REM-sleep trend in mammals. They don't have any. Although there is one case in 1969 suggesting that a pilot whale was in REM sleep for six minutes, most of our assessments to date have not discovered REM sleep—or at least what many sleep scientists would believe to be true REM sleep—in aquatic mammals. From one perspective, this makes sense: when an organism enters REM sleep, the brain paralyzes the body, turning it limp and immobile. Swimming is vital for aquatic mammals, since they must surface to breathe. If full paralysis was to take hold during sleep, they could not swim and would drown.

The mystery deepens when we consider pinnipeds (one of my all-time favorite words, from the Latin derivatives: *pinna* "fin" and *pedis* "foot"), such as fur seals. Partially aquatic mammals, they split their time between land and sea. When on land, they have both NREM sleep and REM sleep, just like humans and all other terrestrial mammals and birds. But when they enter the ocean, they stop having REM sleep almost entirely. Seals in the ocean will sample but a soupçon of the stuff, racking up just 5 to 10 percent of the REM sleep amounts they would normally enjoy when on land. Up to two weeks of ocean-bound time have been documented without any observable REM sleep in seals, who survive in such times on a snooze diet of NREM sleep.

These anomalies do not necessarily challenge the usefulness of REM sleep. Without doubt, REM sleep, and even dreaming, appears to be highly useful and adaptive in those species that have it, as we shall see in part 3 of the book. That REM sleep returns when these animals return to land, rather being done away with entirely, affirms this. It is simply that REM sleep does not appear to be feasible or needed by aquatic mammals when in the ocean. During that time, we assume they make do with NREM sleep—which, for dolphins and whales, may always be the case.

Personally, I don't believe aquatic mammals, even cetaceans like dolphins and whales, have a total absence of REM sleep (though several of my scientific colleagues will tell you I'm wrong). Instead, I think the form of REM sleep these mammals obtain in the ocean is somewhat different and harder to detect: be it brief in nature, occurring at times when we have not been able to observe it, or expressed in ways or hiding in parts of the brain that we have not yet been able to measure.

In defense of my contrarian point of view, I note that it was once believed that egg-laying mammals (monotremes), such as the spiny anteater and the duck-billed platypus, did not have REM sleep. It turned out that they do, or at least a version of it. The outer surface of their brain—the cortex—from which most scientists measure sleeping brainwaves, does not exhibit the choppy, chaotic characteristics of REM-sleep activity. But when scientists looked a little deeper, beautiful bursts of REM-sleep electrical brainwave activity were found at the base of the brain—waves that are a perfect match for those seen in all other mammals. If anything, the duck-billed platypus generates more of this kind of electrical REM-sleep activity than any other mammal! So they did have REM sleep after all, or at least a beta version of it, first rolled out in these more evolutionarily ancient mammals. A fully operational, whole-brain version of REM sleep appears to have been introduced in more developed mammals that later evolved. I believe a similar story of atypical, but nevertheless present, REM sleep will ultimately be observed in dolphins and whales and seals when in the ocean. After all, absence of evidence is not evidence of absence.

More intriguing than the poverty of REM sleep in this aquatic corner of the mammalian kingdom is the fact that birds and mammals evolved separately. REM sleep may therefore have been birthed twice in the course of evolution: once for birds and once for mammals. A common evolutionary pressure may still have created REM sleep in both, in the same way that eyes have evolved separately and independently numerous times across different phyla throughout evolution for the common purpose of visual perception. When a theme repeats in evolution, and independently across unrelated lineages, it often signals a fundamental need.

However, a very recent report has suggested that a proto form of REM sleep exists in an Australian lizard, which, in terms of the evolutionary timeline, predates the emergence of birds and mammals. If this finding is replicated, it would suggest that the original seed of REM sleep was present at least 100 million years earlier than our original estimates. This common seed in certain reptiles may have then germinated into the full form of REM sleep we now see in birds and mammals, including humans.

Regardless of when true REM sleep emerged in evolution, we are fast discovering why REM-sleep dreaming came into being, what vital needs it supports in the warm-blooded world of birds and mammals (e.g., cardiovascular health, emotional restoration, memory association, creativity, body-temperature regulation), and whether other species dream. As we will later discuss, it seems they do.

Setting aside the issue of whether all mammals have REM sleep, an uncontested fact is this: NREM sleep was first to appear in evolution. It is the original form that sleep took when stepping out from behind evolution's creative curtain—a true pioneer. This seniority leads to another intriguing question, and one that I get asked in almost every public lecture I give: Which type of sleep—NREM or REM sleep—is more important? Which do we really *need*?

There are many ways you can define "importance" or "need," and thus numerous ways of answering the question. But perhaps the simplest recipe is to take an organism that has both sleep types, bird or mammal, and keep it awake all night and throughout the subsequent day. NREM and REM sleep are thus similarly removed, creating the conditions of equivalent hunger for each sleep stage. The question is, which type of sleep will the brain feast on when you offer it the chance to consume both during a recovery night? NREM *and* REM sleep in equal proportions? Or more of one than the other, suggesting greater importance of the sleep stage that dominates?

This experiment has now been performed many times on numerous species of birds and mammals, humans included. There are two clear outcomes. First, and of little surprise, sleep duration is far longer on the recovery night (ten or even twelve hours in humans) than during a standard night without prior deprivation (eight hours for us). Responding to the debt, we are essentially trying to "sleep it off," the technical term for which is a sleep rebound.

Second, NREM sleep rebounds harder. The brain will consume a larger portion of deep NREM sleep than of REM sleep on the first night after total sleep deprivation, expressing a lopsided hunger. Despite both sleep types being on offer at the finger buffet of recovery sleep, the brain opts to heap much more deep NREM sleep onto its plate. In the battle of importance, NREM sleep therefore wins. Or does it?

Not quite. Should you keep recording sleep across a second, third, and even fourth recovery night, there's a reversal. Now REM sleep becomes the primary dish of choice with each returning visit to the recovery buffet table, with a side of NREM sleep added. Both sleep stages are therefore essential. We try to recover one (NREM) a little sooner than the other (REM), but the brain will attempt to recoup both, trying to salvage some of the losses incurred. It is important to note, however, that regardless of the amount of recovery opportunity, the brain never comes close to getting back all the sleep it has lost. This is true for total sleep time, just as it is for NREM sleep and for REM sleep. That humans (and all other species) can never "sleep back" that which we have previously lost is one of the most important take-homes of this book, the consequences of which I will describe in chapters 7 and 8.

While you may not be able to repay an accumulated sleep debt in full, a related but very different question has received less attention: can you plan for an upcoming bout of sleep deprivation by "banking" extra sleep credit beforehand? The former speaks to making up a prior debt with later credit. The latter concerns building up credit to deal with a future debt. If the latter turns out to be true, it would be highly applicable in circumstances where upcoming sleep deprivation is predictable, such as emergency services, medicine, aviation and the military.

IF ONLY HUMANS COULD

A third striking difference in sleep across the animal kingdom is the *way* in which we all do it. Here, the diversity is remarkable and, in some cases, almost impossible to believe. Take cetaceans, such as dolphins and whales, for example. Their sleep, of which there is only NREM, can be unihemispheric, meaning they will sleep with half a brain at a time! One half of the brain must always stay awake to maintain life-necessary movement in the aquatic environment. But the other half of the brain will, at times, fall into the most beautiful NREM sleep. Deep, powerful, rhythmic, and slow brainwaves will drench the entirety of one cerebral hemisphere, yet the other half of the cerebrum will be bristling with frenetic, fast brainwave activity, fully awake. This despite the fact that both hemispheres are heavily wired together with thick crisscross fibers, and sit mere millimeters apart, as in human brains.

Of course, both halves of the dolphin brain can be, and frequently are, awake at the very same time, operating in unison. But when it is time for sleep, the two sides of the brain can uncouple and operate independently, one side remaining awake while the other side snoozes away. After this one half of the brain has consumed its fill of sleep, they switch, allowing the previously vigilant half of the brain to enjoy a well-earned period of deep NREM slumber. Even with half of the brain asleep, dolphins can achieve an impressive level of movement and even some vocalized communication.

The neural engineering and tricky architecture required to accomplish this staggering trick of oppositional "lights-on, lights-off" brain activity is rare. Surely Mother Nature could have found a way to avoid sleep entirely under the extreme pressure of nonstop, 24/7 aquatic movement. Would that not have been easier than masterminding a convoluted split-shift system between brain halves for sleep, while still allowing for a joint operating system where both sides unite when awake? Apparently not. Sleep is of such vital necessity that no matter what the evolutionary demands of an organism, even the unyielding need to swim *in perpetuum* from birth to death, Mother Nature had no choice. Sleep with both sides of the brain, or sleep with just one side and then switch. Both are possible, but sleep you must. Sleep is non-negotiable.

The gift of split-brain deep NREM sleep is not entirely unique to aquatic mammals. Birds can do it, too. However, there is a somewhat different, though equally life-preserving, reason: it allows them to keep an eye on things, quite literally. When birds are alone, one half of the brain and its corresponding (opposite-side) eye must stay awake, maintaining vigilance to environmental threats. As it does so, the other eye closes, allowing its corresponding half of the brain to sleep.

Things get even more interesting when birds group together. In some species, many of the birds in a flock will sleep with both halves of the brain at the same time. How do they remain safe from threat? The answer is truly ingenious. The flock will first line up in a row. With the exception of the birds at each end of the line, the rest of the group will allow both halves of the brain to indulge in sleep. Those at the far left and right ends of the row aren't so lucky. They will enter deep sleep with just one half of the brain (opposing in each), leaving the corresponding left and right eye of each bird wide open. In doing so, they provide full panoramic threat detection for the entire group, maximizing the total number of brain halves that can sleep within the flock. At some point, the two end-guards will stand up, rotate 180 degrees, and sit back down, allowing the other side of their respective brains to enter deep sleep.

We mere humans and a select number of other terrestrial mammals appear to be far less skilled than birds and aquatic mammals, unable as we are to take our medicine of NREM sleep in half-brain measure. Or are we?

Two recently published reports suggest humans have a very mild version of unihemispheric sleep—one that is drawn out for similar reasons. If you compare the electrical depth of the deep NREM slow brainwaves on one half of someone's head relative to the other when they are sleeping at home, they are about the same. But if you bring that person into a sleep laboratory, or take them to a hotel—both of which are unfamiliar sleep environments—one half of the brain sleeps a little lighter than the other, as if it's standing guard with just a tad more vigilance due to the potentially less safe context that the conscious brain has registered while awake. The more nights an individual sleeps in the new location, the more similar the sleep is in each half of the brain. It is perhaps one reason why so many of us sleep so poorly the first night in a hotel room.

This phenomenon, however, doesn't come close to the complete division between full wakefulness and truly deep NREM sleep achieved by each side of birds' and dolphins' brains. Humans always have to sleep with both halves of our brain in some state of NREM sleep. Imagine, though, the possibilities that would become available if only we could rest our brains, one half at a time.

I should note that REM sleep is strangely immune to being split across sides of the brain, no matter who you are. All birds, irrespective of the environmental situation, always sleep with both halves of the brain during REM sleep. The same is true for every species that experiences dream sleep, humans included. Whatever the functions of REM-sleep dreaming—and there appear to be many—they require participation of both sides of the brain at the same time, and to an equal degree.

UNDER PRESSURE

The fourth and final difference in sleep across the animal kingdom is the way in which *sleep patterns* can be diminished under rare and very special circumstances, something that the US government sees as a matter of national security, and has spent sizable taxpayer dollars investigating.

The infrequent situation happens only in response to extreme environmental pressures or challenges. Starvation is one example. Place an organism under conditions of severe famine, and foraging for food will supersede sleep. Nourishment will, for a time, push aside the need for sleep, though it cannot be sustained for long. Starve a fly and it will stay awake longer, demonstrating a pattern of food-seeking behavior. The same is true for humans. Individuals who are deliberately fasting will sleep less as the brain is tricked into thinking that food has suddenly become scarce.

Another rare example is the joint sleep deprivation that occurs in female killer whales and their newborn calves. Female killer whales give birth to a single calf once every three to eight years. Calving normally takes place away from the other members of the pod. This leaves the newborn calf incredibly vulnerable during the initial weeks of life, especially during the return to the pod as it swims beside its mother. Up to 50 percent of all new calves are killed during this journey home. It is so dangerous, in fact, that neither mother nor calf appear to sleep while in transit. No mother-calf pair that scientists have observed shows signs of robust sleep en route. This is especially surprising in the calf, since the highest demand and consumption of sleep in every other living species is in the first days and weeks of life, as any new parent will tell you. Such is the egregious peril of long-range ocean travel that these infant whales will reverse an otherwise universal sleep trend.

Yet the most incredible feat of deliberate sleep deprivation belongs to that of birds during transoceanic migration. During this climate-driven race across thousands of miles, entire flocks will fly for many more hours than is normal. As a result, they lose much of the stationary opportunity for plentiful sleep. But even here, the brain has found an ingenious way to obtain sleep. In-flight, migrating birds will grab remarkably brief periods of sleep lasting only seconds in duration. These ultra–power naps are just sufficient to avert the ruinous brain and body deficits that would otherwise ensue from prolonged total sleep deprivation. (If you're wondering, humans have no such similar ability.)

The white-crowned sparrow is an astonishing example of avian sleep deprivation during long-distance flights. This small, quotidian bird is capable of a spectacular feat that the American military has spent millions of research dollars studying. The sparrow has an unparalleled, though time-limited, resilience to total sleep deprivation, one that we humans could never withstand. If you sleep-deprive this sparrow in the laboratory during the migratory period of the year (when it would otherwise be in flight), it suffers virtually no ill effects whatsoever. However, depriving the same sparrow of the same amount of sleep *outside* this migratory time window inflicts a maelstrom of brain and body dysfunction. This humble passerine bird has evolved an extraordinary biological cloak of resilience to total sleep deprivation: one that it deploys only during a time of great survival necessity. You can now imagine why the US government continues to have a vested interest in discovering exactly what that biological suit of armor is: their hope for developing a twenty-four-hour soldier.

HOW SHOULD WE SLEEP?

Humans are not sleeping the way nature intended. The number of sleep bouts, the duration of sleep, and when sleep occurs have all been comprehensively distorted by modernity.

Throughout developed nations, most adults currently sleep in a *monophasic* pattern—that is, we try to take a long, single bout of slumber at night, the average duration of which is now less than seven hours. Visit cultures that are untouched by electricity and you often see something rather different. Hunter-gatherer tribes, such as the Gabra in northern Kenya or the San people in the Kalahari Desert, whose way of life has changed little over the past thousands of years, sleep in a *biphasic* pattern. Both these groups take a similarly longer sleep period at night (seven to eight hours of time in bed, achieving about seven hours of sleep), followed by a thirty- to sixty-minute nap in the afternoon.

There is also evidence for a mix of the two sleep patterns, determined by time of year. Pre-industrial tribes, such as the Hadza in northern Tanzania or the San of Namibia, sleep in a biphasic pattern in the hotter summer months, incorporating a thirty- to forty-minute nap at high noon. They then switch to a largely monophasic sleep pattern during the cooler winter months.

Even when sleeping in a monophasic pattern, the timing of slumber observed in pre-industrialized cultures is not that of our own, contorted making. On average, these tribespeople will fall asleep two to three hours after sunset, around nine p.m. Their nighttime sleep bouts will come to an end just prior to, or soon after, dawn. Have you ever wondered about the meaning of the term "midnight"? It of course means the middle of the night, or, more technically, the middle point of the solar cycle. And so it is for the sleep cycle of hunter-gatherer cultures, and presumably all those that came before. Now consider our cultural sleep norms. Midnight is no longer "mid night." For many of us, midnight is usually the time when we consider checking our email one last time—and we know what often happens in the protracted thereafter. Compounding the problem, we do not then sleep any longer into the morning hours to accommodate these later sleep-onset times. We cannot. Our circadian biology, and the insatiable early-morning demands of a post-industrial way of life, denies us the sleep we vitally need. At one time we went to bed in the hours after dusk and woke up with the chickens. Now many of us are still waking up with the chickens, but dusk is simply the time we are finishing up at the office, with much of the waking night to go. Moreover, few of us enjoy a full afternoon nap, further contributing to our state of sleep bankruptcy.

The practice of biphasic sleep is not cultural in origin, however. It is deeply biological. All humans, irrespective of culture or geographical location, have a genetically hardwired dip in alertness that occurs in the midafternoon hours. Observe any post-lunch meeting around a boardroom table and this fact will become evidently clear. Like puppets whose control strings were let loose, then rapidly pulled taut, heads will start dipping then quickly snap back upright. I'm sure you've experienced this blanket of drowsiness that seems to take hold of you, midafternoon, as though your brain is heading toward an unusually early bedtime.

Both you and the meeting attendees are falling prey to an evolutionarily imprinted lull in wakefulness that favors an afternoon nap, called the post-prandial alertness dip (from the Latin *prandium*, "meal"). This brief descent from high-degree wakefulness to low-level alertness reflects an innate drive to be asleep and napping in the afternoon, and not working. It appears to be a normal part of the daily rhythm of life. Should you ever have to give a presentation at work, for your own sake—and that of the conscious state of your listeners—if you can, avoid the midafternoon slot.

What becomes clearly apparent when you step back from these details is that modern society has divorced us from what should be a preordained arrangement of biphasic sleep—one that our genetic code nevertheless tries to rekindle every afternoon. The separation from biphasic sleep occurred at, or even before, our shift from an agrarian existence to an industrial one.

Anthropological studies of pre-industrial hunter-gatherers have also dispelled a popular myth about how humans should sleep.* Around the close of the early modern era (circa late seventeenth and early eighteenth centuries), historical texts suggest that Western Europeans would take two long bouts of sleep at night, separated by several hours of wakefulness. Nestled in-between these twin slabs of sleep—sometimes called first sleep and second sleep, they would read, write, pray, make love, and even socialize.

This practice may very well have occurred during this moment in human history, in this geographical region. Yet the fact that no pre-industrial cultures studied to date demonstrate a similar nightly split-shift of sleep suggests that it is not the natural, evolutionarily programmed form of human sleep. Rather, it appears to have been a cultural phenomenon that appeared and was popularized with the western European migration. Furthermore, there is no biological rhythm—of brain activity, neurochemical activity, or metabolic activity—that would hint at a human desire to wake up for several hours in the middle of the night. Instead, the true pattern of biphasic sleep—for which there is anthropological, biological, and genetic evidence, and which remains measurable in all human beings to date—is one consisting of a longer bout of continuous sleep at night, followed by a shorter midafternoon nap.

Accepting that this is our natural pattern of slumber, can we ever know for certain what types of health consequences have been caused

*A. Roger Ekirch, *At Day's Close: Night in Times Past* (New York: W. W. Norton, 2006).

by our abandonment of biphasic sleep? Biphasic sleep is still observed in several siesta cultures throughout the world, including regions of South America and Mediterranean Europe. When I was a child in the 1980s, I went on vacation to Greece with my family. As we walked the streets of the major metropolitan Greek cities we visited, there were signs hanging in storefront windows that were very different from those I was used to back in England. They stated: open from nine a.m. to one p.m., closed from one to five p.m., open five to nine p.m.

Today, few of those signs remain in windows of shops throughout Greece. Prior to the turn of the millennium, there was increasing pressure to abandon the siesta-like practice in Greece. A team of researchers from Harvard University's School of Public Health decided to quantify the health consequences of this radical change in more than 23,000 Greek adults, which contained men and women ranging in age from twenty to eighty-three years old. The researchers focused on cardiovascular outcomes, tracking the group across a six-year period as the siesta practice came to an end for many of them.

As with countless Greek tragedies, the end result was heartbreaking, but here in the most serious, literal way. None of the individuals had a history of coronary heart disease or stroke at the start of the study, indicating the absence of cardiovascular ill health. However, those that abandoned regular siestas went on to suffer a 37 percent increased risk of death from heart disease across the six-year period, relative to those who maintained regular daytime naps. The effect was especially strong in workingmen, where the ensuing mortality risk of not napping increased by well over 60 percent.

Apparent from this remarkable study is this fact: when we are cleaved from the innate practice of biphasic sleep, our lives are shortened. It is perhaps unsurprising that in the small enclaves of Greece where siestas still remain intact, such as the island of Ikaria, men are nearly four times as likely to reach the age of ninety as American males. These napping communities have sometimes been described as "the places where people forget to die." From a prescription written long ago in our ancestral genetic code, the practice of natural biphasic sleep, and a healthy diet, appear to be the keys to a long-sustained life.

WE ARE SPECIAL

Sleep, as you can now appreciate, is a unifying feature across the animal kingdom, yet within and between species there is remarkable diversity in amount (e.g., time), form (e.g., half-brain, whole-brain), and pattern (monophasic, biphasic, polyphasic). But are we humans special in our sleep profile, at least, in its pure form when unmolested by modernity? Much has been written about the uniqueness of *Homo sapiens* in other domains—our cognition, creativity, culture, and the size and shape of our brains. Is there anything similarly exceptional about our nightly slumber? If so, could this unique sleep be an unrecognized cause of these aforementioned accomplishments that we prize as so distinctly human—the justification of our hominid name (*Homo sapiens*—Latin derivative, "wise person")?

As it turns out, we humans *are* special when it comes to sleep. Compared to Old- and New-World monkeys, as well as apes, such as chimpanzees, orangutans, and gorillas, human sleep sticks out like the proverbial sore thumb. The total amount of time we spend asleep is markedly shorter than all other primates (eight hours, relative to the ten to fifteen hours of sleep observed in all other primates), yet we have a disproportionate amount of REM sleep, the stage in which we dream. Between 20 and 25 percent of our sleep time is dedicated to REM sleep dreaming, compared to an average of only 9 percent across all other primates! We are the anomalous data point when it comes to sleep time and dream time, relative to all other monkeys and apes. To understand how and why our sleep is so different is to understand the evolution of ape to man, from tree to ground.

Humans are exclusive terrestrial sleepers—we catch our *Z*s lying on the ground (or sometimes raised a little off it, on beds). Other primates will sleep arboreally, on branches or in nests. Only occasionally will other primates come out of trees to sleep on the ground. Great apes, for example, will build an entirely new treetop sleep nest, or platform, every single night. (Imagine having to set aside several hours each evening after dinner to construct a new IKEA bedframe before you can sleep!)

Sleeping in trees was an evolutionarily wise idea, up to a point. It provided safe haven from large, ground-hunting predators, such as hyenas, and small blood-sucking arthropods, including lice, fleas, and ticks.

But when sleeping twenty to fifty feet up in the air, one has to be careful. Become too relaxed in your sleep depth when slouched on a branch or in a nest, and a dangling limb may be all the invitation gravity needs to bring you hurtling down to Earth in a life-ending fall, removing you from the gene pool. This is especially true for the stage of REM sleep, in which the brain completely paralyzes all voluntary muscles of the body, leaving you utterly limp—a literal bag of bones with no tension in your muscles. I'm sure you have never tried to rest a full bag of groceries on a tree branch, but I can assure you it's far from easy. Even if you manage the delicate balancing act for a moment, it doesn't last long. This body-balancing act was the challenge and danger of tree sleeping for our primate forebears, and it markedly constrained their sleep.

Homo erectus, the predecessor of *Homo sapiens*, was the first obligate biped, walking freely upright on two legs. We believe that *Homo erectus* was also the first dedicated ground sleeper. Shorter arms and an upright stance made tree living and sleeping very unlikely. How did *Homo erectus* (and by inference, *Homo sapiens*) survive in the predator-rich ground-sleeping environment, when leopards, hyenas, and saber-toothed tigers (all of which can hunt at night) are on the prowl, and terrestrial bloodsuckers abound? Part of the answer is fire. While there remains some debate, many believe that *Homo erectus* was the first to use fire, and fire was one of the most important catalysts—if not the most important—that enabled us to come out of the trees and live on terra firma. Fire is also one of the best explanations for how we were able to sleep safely on the ground. Fire would deter large carnivores, while the smoke provided an ingenious form of nighttime fumigation, repelling small insects ever keen to bite into our epidermis.

Fire was no perfect solution, however, and ground sleeping would have remained risky. An evolutionary pressure to become qualitatively more efficient in how we sleep therefore developed. Any *Homo erectus* capable of accomplishing more efficient sleep would likely have been favored in survival and selection. Evolution saw to it that our ancient form of sleep became somewhat shorter in *duration*, yet increased in *intensity*, especially by enriching the amount of REM sleep we packed into the night.

In fact, as is so often the case with Mother Nature's brilliance, the problem became part of the solution. In other words, the act of sleeping on solid ground, and not on a precarious tree branch, was the impetus for the enriched and enhanced amounts of REM sleep that developed, while the amount of time spent asleep was able to modestly decrease. When sleeping on the ground, there's no more risk of falling. For the first time in our evolution, hominids could consume all the body-immobilized REM-sleep dreaming they wanted, and not worry about the lasso of gravity whipping them down from treetops. Our sleep therefore became "concentrated": shorter and more consolidated in duration, packed aplenty with high-quality sleep. And not just any type of sleep, but REM sleep that bathed a brain rapidly accelerating in complexity and connectivity. There are species that have more total REM time than hominids, but there are none who power up and lavish such vast proportions of REM sleep onto such a complex, richly interconnected brain as we *Homo sapiens* do.

From these clues, I offer a theorem: the tree-to-ground reengineering of sleep was a key trigger that rocketed *Homo sapiens* to the top of evolution's lofty pyramid. At least two features define human beings relative to other primates. I posit that both have been beneficially and causally shaped by the hand of sleep, and specifically our intense degree of REM sleep relative to all other mammals: (1) our degree of sociocultural complexity, and (2) our cognitive intelligence. REM sleep, and the act of dreaming itself, lubricates both of these human traits.

To the first of these points, we have discovered that REM sleep exquisitely recalibrates and fine-tunes the emotional circuits of the human brain (discussed in detail in part 3 of the book). In this capacity, REM sleep may very well have accelerated the richness and rational control of our initially primitive emotions, a shift that I propose critically contributed to the rapid rise of *Homo sapiens* to dominance over all other species in key ways.

We know, for example, that REM sleep increases our ability to recognize and therefore successfully navigate the kaleidoscope of socioemotional signals that are abundant in human culture, such as overt and covert facial expressions, major and minor bodily gestures, and even mass group behavior. One only needs to consider disorders such as autism to see how challenging and different a social existence can be without these emotional navigation abilities being fully intact.

Related, the REM-sleep gift of facilitating accurate recognition and comprehension allows us to make more intelligent decisions and actions as a consequence. More specifically, the coolheaded ability to regulate our emotions each day—a key to what we call emotional IQ—depends on getting sufficient REM sleep night after night. (If your mind immediately jumped to particular colleagues, friends, and public figures who lack these traits, you may well wonder about how much sleep, especially late-morning REM-rich sleep, they are getting.)

Second, and more critical, if you multiply these individual benefits within and across groups and tribes, all of which are experiencing an ever-increasing intensity and richness of REM sleep over millennia, we can start to see how this nightly REM-sleep recalibration of our emotional brains could have scaled rapidly and exponentially. From this REM-sleep-enhanced emotional IQ emerged a new and far more sophisticated form of hominid socioecology across vast collectives, one that helped enable the creation of large, emotionally astute, stable, highly bonded, and intensely social communities of humans.

I will go a step further and suggest that this is *the* most influential function of REM sleep in mammals, perhaps the most influential function of *all* types of sleep in *all* mammals, and even the most eminent advantage ever gifted by sleep in the annals of all planetary life. The adaptive benefits conferred by complex emotional processing are truly monumental, and so often overlooked. We humans can instantiate vast numbers of emotions in our embodied brains, and thereafter, deeply experience and even regulate those emotions. Moreover, we can recognize and help shape the emotions of others. Through both of these intra- and interpersonal processes, we can forge the types of cooperative alliances that are necessary to establish large social groups, and beyond groups, entire societies brimming with powerful structures and ideologies. What may at first blush have seemed like a modest asset awarded by REM sleep to a single *individual* is, I believe, one of the most valuable commodities ensuring the survival and dominance of our species as a *collective.*

The second evolutionary contribution that the REM-sleep dreaming state fuels is creativity. NREM sleep helps transfer and make safe newly learned information into long-term storage sites of the brain. But it is REM sleep that takes these freshly minted memories and begins colliding them with the entire back catalog of your life's autobiography. These mnemonic collisions during REM sleep spark new creative insights as novel links are forged between unrelated pieces of information. Sleep cycle by sleep cycle, REM sleep helps construct vast associative networks of information within the brain. REM sleep can even take a step back, so to speak, and divine overarching insights and gist: something akin to general knowledge—that is, what a collection of information means as a whole, not just an inert back catalogue of facts. We can awake the next morning with new solutions to previously intractable problems or even be infused with radically new and original ideas.

Adding, then, to the opulent and domineering socioemotional fabric that REM sleep helps weave across the masses came this second, creativity benefit of dream sleep. We should (cautiously) revere how superior our hominid ingenuity is relative to that of any of our closest rivals, primate or other. The chimpanzees—our nearest living primate relatives—have been around approximately 5 million years longer than we have; some of the great apes preceded us by at least 10 million years. Despite aeons of opportunity time, neither species has visited the moon, created computers, or developed vaccines. Humbly, we humans have. Sleep, especially REM sleep and the act of dreaming, is a tenable, yet underappreciated, factor underlying many elements that form our unique human ingenuity and accomplishments, just as much as language or tool use (indeed, there is even evidence that sleep causally shapes both these latter traits as well).

Nevertheless, the superior emotional brain gifts that REM sleep affords should be considered more influential in defining our hominid success than the second benefit, of inspiring creativity. Creativity is an evolutionarily powerful tool, yes. But it is largely limited to an individual. Unless creative, ingenious solutions can be shared between individuals through the emotionally rich, pro-social bonds and cooperative relationships that REM sleep fosters—then creativity is far more likely to remain fixed within an individual, rather than spread to the masses.

Now we can appreciate what I believe to be a classic, self-fulfilling positive cycle of evolution. Our shift from tree to ground sleeping instigated an ever more bountiful amount of relative REM sleep compared with other primates, and from this bounty emerged a steep increase in cognitive creativity, emotional intelligence, and thus social complexity. This, alongside our increasingly dense, interconnected brains, led to improved daily (and nightly) survival strategies. In turn, the harder we worked those increasingly developed emotional and creative circuits of the brain during the day, the greater was our need to service and recalibrate these ever-demanding neural systems at night with more REM sleep.

As this positive feedback loop took hold in exponential fashion, we formed, organized, maintained, and deliberatively shaped ever larger social groups. The rapidly increasing creative abilities could thus be spread more efficiently and rapidly, and even improved by that ever-increasing amount of hominid REM-sleep that enhances emotional and social sophistication. REM-sleep dreaming therefore represents a tenable new contributing factor, among others, that led to our astonishingly rapid evolutionary rise to power, for better and worse—a new (sleep-fueled), globally dominant *social* superclass.

CHAPTER 5

Changes in Sleep Across the Life Span

SLEEP BEFORE BIRTH

Through speech or song, expecting parents will often thrill at their ability to elicit small kicks and movements from their in utero child. Though you should never tell them this, the baby is most likely fast asleep. Prior to birth, a human infant will spend almost all of its time in a sleep-like state, much of which resembles the REM-sleep state. The sleeping fetus is therefore unaware of its parents' performative machinations. Any co-occurring arm flicks and leg bops that the mother feels from her baby are most likely to be the consequence of random bursts of brain activity that typify REM sleep.

Adults do not—or at least should not—throw out similar nighttime kicks and movements, since they are held back by the body-paralyzing mechanism of REM sleep. But in utero, the immature fetus's brain has yet to construct the REM-sleep muscle-inhibiting system adults have in place. Other deep centers of the fetus brain have, however, already been glued in place, including those that generate sleep. Indeed, by the end of the second trimester of development (approximately week 23 of pregnancy), the vast majority of the neural dials and switches required to produce NREM and REM sleep have been sculpted out and wired up. As a result of this mismatch, the fetus brain still generates formidable motor commands during REM sleep, except there is no paralysis to hold them back. Without restraint, those commands are freely translated into frenetic body movements, felt by the mother as acrobatic kicks and featherweight punches.

At this stage of in utero development, most of the time is spent in sleep. The twenty-four-hour period contains a mishmash of approximately six hours of NREM sleep, six hours of REM sleep, and twelve hours of an intermediary sleep state that we cannot confidently say is REM or NREM sleep, but certainly is not full wakefulness. It is only when the fetus enters the final trimester that the glimmers of real wakefulness emerge. Far less than you would probably imagine, though—just two to three hours of each day are spent awake in the womb.

Even though total sleep time decreases in the last trimester, a paradoxical and quite ballistic increase in REM-sleep time occurs. In the last two weeks of pregnancy, the fetus will ramp up its consumption of REM sleep to almost nine hours a day. In the last week before birth, REM-sleep amount hits a lifetime high of twelve hours a day. With near insatiable appetite, the human fetus therefore ramps up its hunger for REM sleep just before entering the world. There will be no other moment during the life of that individual—pre-natal, early post-natal, adolescence, adulthood, or old age—when they will undergo such a dramatic change in REM-sleep need, or feast so richly on the stuff.

Is the fetus actually dreaming when in REM sleep? Probably not in the way most of us conceptualize dreams. But we do know that REM sleep is vital for promoting brain maturation. The construction of a human being in the womb occurs in distinct, interdependent stages, a little bit like building a house. You cannot crown a house with a roof before there are supporting wall frames to rest it on, and you cannot put up walls without a foundation to seat them in. The brain, like the roof of a house, is one of the last items to be constructed during development. And like a roof, there are sub-stages to that process—you need a roof frame before you can start adding roof tiles, for instance.

Detailed creation of the brain and its component parts occurs at a rapid pace during the second and third trimesters of human development—precisely the time window when REM-sleep amounts skyrocket. This is no coincidence. REM sleep acts as an electrical fertilizer during this critical phase of early life. Dazzling bursts of electrical activity during REM sleep stimulate the lush growth of neural pathways all over the developing brain, and then furnish each with a healthy bouquet of connecting ends, or synaptic terminals. Think of REM sleep like an Internet service provider that populates new neighborhoods of the brain with vast networks of fiber-optic cables. Using these inaugural bolts of electricity, REM sleep then activates their high-speed functioning.

This phase of development, which infuses the brain with masses of neural connections, is called *synaptogenesis*, as it involves the creation of millions of wiring links, or synapses, between neurons. By deliberate design, it is an overenthusiastic first pass at setting up the mainframe of a brain. There is a great deal of redundancy, offering many, many possible circuit configurations to emerge within the infant's brain once born. From the perspective of the Internet service provider analogy, all homes, across all neighborhoods, throughout all territories of the brain have been gifted a high degree of connectivity and bandwidth in this first phase of life.

Charged with such a herculean task of neuro-architecture—establishing the neural highways and side streets that will engender thoughts, memories, feelings, decisions, and actions—it's no wonder REM sleep must dominate most, if not all, of early developmental life. In fact, this is true for all other mammals:* the time of life when REM sleep is greatest is the same stage when the brain is undergoing the greatest construction.

Worryingly, if you disturb or impair the REM sleep of a developing infant brain, pre- or early post-term, there are consequences. In the 1990s, researchers began studying newly born rat pups. Simply by blocking REM sleep, their gestational progress was retarded, despite chronological time marching on. The two should, of course, progress in unison. Depriving the infant rats of REM sleep stalled construction of their neural rooftop—the cerebral cortex of the brain. Without REM sleep, assembly work on the brain ground to a halt, frozen in time by the experimental wedge of a lack of REM sleep. Day after day, the half-finished roofline of the sleep-starved cerebral cortex shows no growth change.

The very same effect has now been demonstrated in numerous

*The exception, noted in chapter 4, may be newborn killer whales. They do not appear to have the chance for sleep right after birth, as they have to make the perilous journey back to their pod from the calving fields miles away, shadowed by their mother. However, this is an assumption. It remains possible that they, like all other mammals, still consume in utero large volumes of sleep, and even REM sleep, just prior to birth. We simply do not yet know.

other mammalian species, suggesting that the effect is probably common across mammals. When the infant rat pups were finally allowed to get some REM sleep, assembly of the cerebral rooftop did restart, but it didn't accelerate, nor did it ever fully get back on track. An infant brain without sleep will be a brain ever underconstructed.

A more recent link with deficient REM sleep concerns autism spectrum disorder (ASD) (not to be confused with attention deficit hyperactivity disorder [ADHD], which we will discuss later in the book). Autism, of which there are several forms, is a neurological condition that emerges early in development, usually around two or three years of age. The core symptom of autism is a lack of social interaction. Individuals with autism do not communicate or engage with other people easily, or typically.

Our current understanding of what causes autism is incomplete, but central to the condition appears to be an inappropriate wiring up of the brain during early developmental life, specifically in the formation and number of synapses—that is, abnormal synaptogenesis. Imbalances in synaptic connections are common in autistic individuals: excess amounts of connectivity in some parts of the brain, deficiencies in others.

Realizing this, scientists have begun to examine whether the sleep of individuals with autism is atypical. It is. Infants and young children who show signs of autism, or who are diagnosed with autism, do not have normal sleep patterns or amounts. The circadian rhythms of autistic children are also weaker than their non-autistic counterparts, showing a flatter profile of melatonin across the twenty-four-hour period rather than a powerful rise in concentration at night and rapid fall throughout the day.* Biologically, it is as if the day and night are far less light and dark, respectively, for autistic individuals. As a consequence, there is a weaker signal for when stable wake and solid sleep should take place. Additionally, and perhaps related, the total amount of sleep that autistic children can generate is less than that of non-autistic children.

---

*S. Cohen, R. Conduit, S. W. Lockley, S. M. Rajaratnam, and K. M. Cornish, "The relationship between sleep and behavior in autism spectrum disorder (ASD): a review," *Journal of Neurodevelopmental Disorders* 6, no. 1 (2011): 44.

Most notable, however, is the significant shortage of REM sleep. Autistic individuals show a 30 to 50 percent deficit in the amount of REM sleep they obtain, relative to children without autism.* Considering the role of REM sleep in establishing the balanced mass of synaptic connections within the developing brain, there is now keen interest in discovering whether or not REM-sleep deficiency is a contributing factor to autism.

Existing evidence in humans is simply correlational, however. Just because autism and REM-sleep abnormalities go hand in hand does not mean that one causes the other. Nor does this association tell you the direction of causality even if it does exist: Is deficient REM sleep causing autism, or is it the other way around? It is curious to note, however, that selectively depriving an infant rat of REM sleep leads to aberrant patterns of neural connectivity, or synaptogenesis, in the brain.† Moreover, rats deprived of REM sleep during infancy go on to become socially withdrawn and isolated as adolescents and adults.‡ Irrespective of causality issues, tracking sleep abnormalities represents a new diagnostic hope for the early detection of autism.

Of course, no expecting mother has to worry about scientists disrupting the REM sleep of their developing fetus. But alcohol can inflict that same selective removal of REM sleep. Alcohol is one of the most powerful suppressors of REM sleep that we know of. We will discuss the reason that alcohol blocks REM-sleep generation, and the consequences of that sleep disruption in adults, in later chapters. For now, however, we'll focus on the impact of alcohol on the sleep of a developing fetus and newborn.

Alcohol consumed by a mother readily crosses the placental barrier, and therefore readily infuses her developing fetus. Knowing this, scientists first examined the extreme scenario: mothers who were alcoholics

---

*A. W. Buckley, A. J. Rodriguez, A. Jennison, et al. "Rapid eye movement sleep percentage in children with autism compared with children with developmental delay and typical development," *Archives of Pediatrics and Adolescent Medicine* 164, no. 11 (2010): 1032–37. See also S. Miano, O. Bruni, M. Elia, A. Trovato, et al., "Sleep in children with autistic spectrum disorder: a questionnaire and polysomnographic study," *Sleep Medicine* 9, no. 1 (2007): 64–70.

†G. Vogel and M. Hagler, "Effects of neonatally administered iprindole on adult behaviors of rats," *Pharmacology Biochemistry and Behavior* 55, no. 1 (1996): 157–61.

‡Ibid.

or heavy drinkers during pregnancy. Soon after birth, the sleep of these neonates was assessed using electrodes gently placed on the head. The newborns of heavy-drinking mothers spent far less time in the active state of REM sleep compared with infants of similar age but who were born of mothers who did not drink during pregnancy.

The recording electrodes went on to point out an even more concerning physiological story. Newborns of heavy-drinking mothers did not have the same electrical quality of REM sleep. You will remember from chapter 3 that REM sleep is exemplified by delightfully chaotic—or desynchronized—brainwaves: a vivacious and healthy form of electrical activity. However, the infants of heavy-drinking mothers emitted a brainwave pattern that was more sedentary in this regard.*

Fortunately, most mothers these days do not drink heavily during pregnancy. But what about the more common situation of an expectant mom having an occasional glass or two of wine during pregnancy? Using noninvasive tracking of heart rate, together with ultrasound measures of body, eye, and breathing movement, we are now able to determine the basic stages of NREM sleep and REM sleep of a fetus when it is in the womb. Equipped with these methods, a group of researchers studied the sleep of babies who were just weeks away from being born. Their mothers were assessed on two successive days. On one of those days, the mothers drank non-alcoholic fluids. On the other day, they drank approximately two glasses of wine (the absolute amount was controlled on the basis of their body weight). Alcohol significantly reduced the amount of time that the unborn babies spent in REM sleep, relative to the non-alcohol condition.

That alcohol also dampened the intensity of REM sleep experienced by the fetus, defined by the standard measure of how many darting rapid eye movements adorn the REM-sleep cycle. Furthermore, these unborn infants suffered a depression in breathing during REM sleep, with breath rates dropping from a rate of 381 per hour

*V. Havlicek, R. Childiaeva, and V. Chernick, "EEG frequency spectrum characteristics of sleep states in infants of alcoholic mothers," *Neuropädiatrie* 8, no. 4 (1977): 360–73. See also S. Loffe, R. Childiaeva, and V. Chernick, "Prolonged effects of maternal alcohol ingestion on the neonatal electroencephalogram," *Pediatrics* 74, no. 3 (1984): 330–35.

during natural sleep to just 4 per hour when the fetus was exposed to alcohol.*

Beyond alcohol abstinence during pregnancy, the time window of nursing also warrants mention. Almost half of all lactating women in Western countries consume alcohol in the months during breastfeeding. Alcohol is readily absorbed in a mother's milk. Concentrations of alcohol in breast milk closely resemble those in a mother's bloodstream: a 0.08 blood alcohol level in a mother will result in approximately a 0.08 alcohol level in breast milk.† Recently we have discovered what alcohol in breast milk does to the sleep of an infant.

Newborns will normally transition straight into REM sleep after a feeding. Many mothers already know this: almost as soon as suckling stops, and sometimes even before, the infant's eyelids will close, and underneath, the eyes will begin darting left-right, indicating that their baby is now being nourished by REM sleep. A once-common myth was that babies sleep better if the mother has had an alcoholic drink before a feeding—beer was the suggested choice of beverage in this old tale. For those of you who are beer lovers, unfortunately, it is just that—a myth. Several studies have fed infants breast milk containing either a non-alcoholic flavor, such as vanilla, or a controlled amount of alcohol (the equivalent of a mother having a drink or two). When babies consume alcohol-laced milk, their sleep is more fragmented, they spend more time awake, and they suffer a 20 to 30 percent suppression of REM sleep after.‡ Often, the babies will even try to get back some of that missing REM sleep once they have cleared it from their bloodstream.

What emerges from all of these studies is that REM sleep is not optional during early human life, but obligatory. Every hour of REM sleep appears to count, as evidenced by the attempt by a fetus or new-

---

*E. J. Mulder, L. P. Morssink, T. van der Schee, and G. H. Visser, "Acute maternal alcohol consumption disrupts behavioral state organization in the near-term fetus," *Pediatric Research* 44, no. 5 (1998): 774–79.

†Beyond sleep, alcohol also inhibits the milk ejection reflex and causes a temporary decrease in milk yield.

‡J. A. Mennella and P. L. Garcia-Gomez, "Sleep disturbances after acute exposure to alcohol in mothers' milk," *Alcohol* 25, no. 3 (2001): 153–58. See also J. A. Mennella and C. J. Gerrish, "Effects of exposure to alcohol in mother's milk on infant sleep," *Pediatrics* 101, no. 5 (1998): E2.

born to regain REM sleep when it is lost.* Sadly, we do not yet fully understand what the long-term effects are of fetal or neonate REM-sleep disruption, alcohol-triggered or otherwise. Only that blocking or reducing REM sleep in newborn animals hinders and distorts brain development, leading to an adult that is socially abnormal.

CHILDHOOD SLEEP

Perhaps the most obvious and tormenting (for new parents) difference between the sleep of infants and young children and that of adults is the number of slumber phases. In contrast to the single, monophasic sleep pattern observed in adults of industrialized nations, infants and young kids display polyphasic sleep: many short snippets of sleep through the day and night, punctuated by numerous awakenings, often vocal.

There is no better or more humorous affirmation of this fact than the short book of lullabies, written by Adam Mansbach, entitled *Go the F**k to Sleep*. Obviously, it's an adult book. At the time of writing, Mansbach was a new father. And like many a new parent, he was run ragged by the constant awakenings of his child: the polyphasic profile of infant sleep. The incessant need to attend to his young daughter, helping her fall back to sleep time and time and time again, night after night after night, left him utterly exasperated. It got to the point where Mansbach just had to vent all the loving rage he had pent up. What came spilling out onto the page was a comedic splash of rhymes he would fictitiously read to his daughter, the themes of which will immediately resonate with many new parents. "I'll read you one very last book if you swear,/You'll go the fuck to sleep." (I implore you to listen to the audiobook version of the work, narrated to perfection by the sensational actor Samuel L. Jackson.)

Fortunately, for all new parents (Mansbach included), the older a child gets, the fewer, longer, and more stable their sleep bouts become.†

*While not directly related to sleep quantity or quality, alcohol use by the mother before co-sleeping with their newborn infants (bed to couch) leads to a seven- to ninefold increase of sudden infant death syndrome (SIDS), compared with those who do not use alcohol. (P. S. Blair, P. Sidebotham, C. Evason-Coombe, et al., "Hazardous cosleeping environments and risk factors amenable to change: case-control study of SIDS in southwest England," *BMJ* 339 [2009]: b3666.)

†The ability for infants and young children to become independent nighttime sleepers

Explaining this change is the circadian rhythm. While the brain areas that generate sleep are molded in place well before birth, the master twenty-four-hour clock that controls the circadian rhythm—the suprachiasmatic nucleus—takes considerable time to develop. Not until age three or four months will a newborn show modest signs of being governed by a daily rhythm. Slowly, the suprachiasmatic nucleus begins to latch on to repeating signals, such as daylight, temperature change, and feedings (so long as those feedings are highly structured), establishing a stronger twenty-four-hour rhythm.

By the one-year milestone of development, the suprachiasmatic nucleus clock of an infant has gripped the steering reins of the circadian rhythm. This means that the child now spends more of the day awake, interspersed with several naps and, mercifully, more of the night asleep. Mostly gone are the indiscriminate bouts of sleep and wake that once peppered the day and night. By four years of age, the circadian rhythm is in dominant command of a child's sleep behavior, with a lengthy slab of nighttime sleep, usually supplemented by just a single daytime nap. At this stage, the child has transitioned from a polyphasic sleep pattern to a biphasic sleep pattern. Come late childhood, the modern, monophasic pattern of sleep is finally made real.

What this progressive establishment of stable rhythmicity hides, however, is a much more tumultuous power struggle between NREM and REM sleep. Although the amount of total sleep gradually declines from birth onwards, all the while becoming more stable and consolidated, the ratio of time spent in NREM sleep and REM sleep does not decline in a similarly stable manner.

During the fourteen hours of total shut-eye per day that a six-month-old infant obtains, there is a 50/50 timeshare between NREM and REM sleep. A five-year-old, however, will have a 70/30 split between NREM and REM sleep across the eleven hours of total daily slumber. In other words, the proportion of REM sleep *decreases* in early childhood

---

is the keen focus of—or perhaps better phrased, the outright obsession of—many new parents. There are innumerable books whose sole focus is to outline the best practices for infant and child sleep. This book is not meant to offer an overview of the topic. If you are looking for information on this issue, I would recommend the pediatric sleep doctor, Craig Canapari. Craig is the Director of the Yale Pediatric Sleep Center and has superb resources for parents on his own website.

while the proportion of NREM sleep actually *increases*, even though total sleep time decreases. The downgrading of the REM-sleep portion, and the upswing in NREM-sleep dominance, continues, throughout early and midchildhood. That balance will finally stabilize to an 80/20 NREM/REM sleep split by the late teen years, and remain so throughout early and midadulthood.

SLEEP AND ADOLESCENCE

Why do we spend so much time in REM sleep in the womb and early in life, yet switch to a heavier dominance of deep NREM sleep in late childhood and early adolescence? If we quantify the intensity of the deep-sleep brainwaves, we see the very same pattern: a decline in REM-sleep intensity in the first year of life, yet an exponential rise in deep NREM sleep intensity in mid- and late childhood, hitting a peak just before puberty, and then damping back down. What's so special about this type of deep sleep at this transitional time of life?

Prior to birth, and soon after, the challenge for development was to build and add vast numbers of neural highways and interconnections that become a fledgling brain. As we have discussed, REM sleep plays an essential role in this proliferation process, helping to populate brain neighborhoods with neural connectivity, and then activate those pathways with a healthy dose of informational bandwidth.

But since this first round of brain wiring is purposefully overzealous, a second round of remodeling must take place. It does so during late childhood and adolescence. Here, the architectural goal is not to scale up, but to scale back for the goal of efficiency and effectiveness. The time of adding brain connections with the help of REM sleep is over. Instead, pruning of connections becomes the order of the day or, should I say, night. Enter the sculpting hand of deep NREM sleep.

Our analogy of the Internet service provider is a helpful one to return to. When first setting up the network, each home in the newly built neighborhood was given an equal amount of connectivity bandwidth and thus potential for use. However, that's an inefficient solution for the long term, since some of these homes will become heavy bandwidth users over time, while other homes will consume very little. Some homes may even remain vacant and never use any bandwidth. To reliably estimate what pattern of demand exists, the Internet service provider needs time to gather usage statistics. Only after a period of experience can the provider make an informed decision on how to refine the original network structure it put in place, dialing back connectivity to low-use homes, while increasing connectivity to other homes with high bandwidth demand. It is not a complete redo of the network, and much of the original structure will remain in place. After all, the Internet service provider has done this many times before, and they have a reasonable estimate of how to build a first pass of the network. But a use-dependent reshaping and downsizing must still occur if maximum network efficiency is to be achieved.

The human brain undergoes a similar, use-determined transformation during late childhood and adolescence. Much of the original structure laid down early in life will persist, since Mother Nature has, by now, learned to create a quite accurate first-pass wiring of a brain after billions of attempts over many thousands of years of evolution. But she wisely leaves something on the table in her generic brain sculpture, that of individualized refinement. The unique experiences of a child during their formative years translate to a set of personal usage statistics. Those experiences, or those statistics, provide the bespoke blueprint for a last round of brain refinement,* capitalizing on the opportunity left open by nature. A (somewhat) generic brain becomes ever more individualized, based on the personalized use of the owner.

To help with the job of refinement and downscaling of connectivity, the brain employs the services of deep NREM sleep. Of the many functions carried out by deep NREM sleep—the full roster of which we will discuss in the next chapter—it is that of synaptic pruning that features prominently during adolescence. In a remarkable series of experiments, the pioneering sleep researcher Irwin Feinberg discovered something fascinating about how this operation of downscaling takes place within the adolescent brain. His findings help justify an opinion you may also hold: adolescents have a less rational version of an adult brain, one that takes more risks and has relatively poor decision-making skills.

---

*Even though the degree of neural network connectivity decreases during development, the physical size of our brain cells, and thus the physical size of the brain and head, increases.

Using electrodes placed all over the head—front and back, left side and right, Feinberg began recording the sleep of a large group of kids starting at age six to eight years old. Every six to twelve months, he would bring these individuals back to his laboratory and perform another sleep measurement. He didn't stop for ten years. He amassed more than 3,500 all-night assessments: a scarcely believable 320,000 hours of sleep recordings! From these, Feinberg created a series of snapshots, depicting how deep-sleep intensity changed with the stages of brain development as the children made their often awkward transition through adolescence into adulthood. It was the neuroscience equivalent of time-lapse photography in nature: taking repeat pictures of a tree as it first comes into bud in the spring (babyhood), then bursts into leaf during the summer (late childhood), then matures in color come the fall (early adolescence), and finally sheds its leaves in the winter (late adolescence and early adulthood).

During mid- and late childhood, Feinberg observed moderate deep-sleep amounts as the last neural growth spurts inside the brain were being completed, analogous to late spring and early summer. Then Feinberg began seeing a sharp rise in deep-sleep intensity in his electrical recordings, right at the time when the developmental needs of brain connectivity switch from growing connections to shedding them; the tree's equivalent of fall. Just as maturational fall was about to turn to winter, and the shedding was nearly complete, Feinberg's recordings showed a clear ramping back down in deep NREM-sleep intensity to lower intensity once more. The life cycle of childhood was over, and as the last leaves dropped, the onward neural passage of these teenagers had been secured. Deep NREM sleep had aided their transition into early adulthood.

Feinberg proposed that the rise and fall of deep-sleep intensity were helping lead the maturational journey through the precarious heights of adolescence, followed by safe onward passage into adulthood. Recent findings have supported his theory. As deep NREM sleep performs its final overhaul and refinement of the brain during adolescence, cognitive skills, reasoning, and critical thinking start to improve, and do so in a proportional manner with that NREM sleep change. Taking a closer look at the timing of this relationship, you see something even more interesting. The changes in deep NREM sleep always precede the cognitive and developmental milestones within the brain by several weeks or months, implying a direction of influence: *deep sleep may be a driving force of brain maturation, not the other way around.*

Feinberg made a second seminal discovery. When he examined the timeline of changing deep-sleep intensity at each different electrode spot on the head, it was not the same. Instead, the rise-and-fall pattern of maturation always began at the back of the brain, which performs the functions of visual and spatial perception, and then progressed steadily forward as adolescence progressed. Most striking, the very last stop on the maturational journey was the tip of the frontal lobe, which enables rational thinking and critical decision-making. Therefore, the back of the brain of an adolescent was more adult-like, while the front of the brain remained more child-like at any one moment during this developmental window of time.*

His findings helped explain why rationality is one of the last things to flourish in teenagers, as it is the last brain territory to receive sleep's maturational treatment. Certainly sleep is not the only factor in the ripening of the brain, but it appears to be a significant one that paves the way to mature thinking and reasoning ability. Feinberg's study reminds me of a billboard advertisement I once saw from a large insurance firm, which read: "Why do most 16-year-olds drive like they're missing part of their brain? Because they are." It takes deep sleep, and developmental time, to accomplish the neural maturation that plugs this brain "gap" within the frontal lobe. When your children finally reach their mid-twenties and your car insurance premium drops, you can thank sleep for the savings.

The relationship between deep-sleep intensity and brain maturation

*With all this talk of removing synapses in the adolescent brain, I should point out that plenty of strengthening continues to occur in the adolescent (and adult) brain within those circuits that remain, and this is carried out by different sleeping brainwaves we'll discuss in the next chapter. Suffice it to say that the ability to learn, retain, and thus remember new memories persists, even when set against the backdrop of general connectivity down-scaling throughout late development. Nevertheless, by teenage years, the brain is less malleable, or plastic, than during infancy or early childhood—one example being the ease with which younger children can pick up a second language compared with older adolescents.

that Feinberg described has now been observed in many different populations of children and adolescents around the world. But how can we be sure that deep sleep truly offers a neural pruning service necessary for brain maturation? Perhaps changes in sleep and brain maturation simply occur at roughly the same time but are independent of each other?

The answer is found in studies of juvenile rats and cats at the equivalent stage to human adolescence. Scientists deprived these animals of deep sleep. In doing so, they halted the maturational refinement of brain connectivity, demonstrating a causal role for deep NREM sleep in propelling the brain into healthy adulthood.* Of concern is that administering caffeine to juvenile rats will also disrupt deep NREM sleep and, as a consequence, delay numerous measures of brain maturation and the development of social activity, independent grooming, and the exploration of the environment—measures of self-motivated learning.†

Recognizing the importance of deep NREM sleep in teenagers has been instrumental to our understanding of healthy development, but it has also offered clues as to what happens when things go wrong in the context of abnormal development. Many of the major psychiatric disorders, such as schizophrenia, bipolar disorder, major depression, and ADHD are now considered disorders of abnormal development, since they commonly emerge during childhood and adolescence.

We will return to the issue of sleep and psychiatric illness several times in the course of this book, but schizophrenia deserves special mention at this juncture. Several studies have tracked neural development using brain scans every couple of months in hundreds of young teenagers as they make their way through adolescence. A proportion of these individuals went on to develop schizophrenia in their late teenage years and early adulthood. Those individuals who developed schizophrenia had an abnormal pattern of brain maturation that was associated with synaptic pruning, especially in the frontal lobe regions where rational, logical thoughts are controlled—the inability to do so

---

*M. G. Frank, N. P. Issa, and M. P. Stryker, "Sleep enhances plasticity in the developing visual cortex," *Neuron* 30, no. 1 (2001): 275–87.

†N. Olini, S. Kurth, and R. Huber, "The effects of caffeine on sleep and maturational markers in the rat," *PLOS ONE* 8, no. 9 (2013): e72539.

being a major symptom of schizophrenia. In a separate series of studies, we have also observed that in young individuals who are at high risk of developing schizophrenia, and in teenagers and young adults with schizophrenia, there is a two- to threefold reduction in deep NREM sleep.* Furthermore, the electrical brainwaves of NREM sleep are not normal in their shape or number in the affected individuals. Faulty pruning of brain connections in schizophrenia caused by sleep abnormalities is now one of the most active and exciting areas of investigation in psychiatric illness.†

Adolescents face two other harmful challenges in their struggle to obtain sufficient sleep as their brains continue to develop. The first is a change in their circadian rhythm. The second is early school start times. I will address the harmful and life-threatening effects of the latter in a later chapter; however, the complications of early school start times are inextricably linked with the first issue—a shift in circadian rhythm. As young children, we often wished to stay up late so we could watch television, or engage with parents and older siblings in whatever it was that they were doing at night. But when given that chance, sleep would usually get the better of us, on the couch, in a chair, or sometimes flat out on the floor. We'd be carried to bed, slumbering and unaware, by those older siblings or parents who could stay awake. The reason is not simply that children need more sleep than their older siblings or parents, but also that the circadian rhythm of a young child runs on an earlier schedule. Children therefore become sleepy earlier and wake up earlier than their adult parents.

Adolescent teenagers, however, have a different circadian rhythm from their young siblings. During puberty, the timing of the suprachiasmatic nucleus is shifted progressively forward: a change that is common across all adolescents, irrespective of culture or geography. So far forward, in fact, it passes even the timing of their adult parents.

As a nine-year-old, the circadian rhythm would have the child asleep

---

*S. Sarkar, M. Z. Katshu, S. H. Nizamie, and S. K. Praharaj, "Slow wave sleep deficits as a trait marker in patients with schizophrenia," *Schizophrenia Research* 124, no. 1 (2010): 127–33.

†M. F. Profitt, S. Deurveilher, G. S. Robertson, B. Rusak, and K. Semba, "Disruptions of sleep/wake patterns in the stable tubule only polypeptide (STOP) null mouse model of schizophrenia," *Schizophrenia Bulletin* 42, no. 5 (2016): 1207–15.

by around nine p.m., driven in part by the rising tide of melatonin at this time in children. By the time that same individual has reached sixteen years of age, their circadian rhythm has undergone a dramatic shift forward in its cycling phase. The rising tide of melatonin, and the instruction of darkness and sleep, is many hours away. As a consequence, the sixteen-year-old will usually have no interest in sleeping at nine p.m. Instead, peak *wakefulness* is usually still in play at that hour. By the time the parents are getting tired, as their circadian rhythms take a downturn and melatonin release instructs sleep—perhaps around ten or eleven p.m., their teenager can still be wide awake. A few more hours must pass before the circadian rhythm of a teenage brain begins to shut down alertness and allow for easy, sound sleep to begin.

This, of course, leads to much angst and frustration for all parties involved on the back end of sleep. Parents want their teenager to be awake at a "reasonable" hour of the morning. Teenagers, on the other hand, having only been capable of initiating sleep some hours after their parents, can still be in their trough of the circadian downswing. Like an animal prematurely wrenched out of hibernation too early, the adolescent brain still needs more sleep and more time to complete the circadian cycle before it can operate efficiently, without grogginess.

If this remains perplexing to parents, a different way to frame and perhaps appreciate the mismatch is this: asking your teenage son or daughter to go to bed and fall asleep at ten p.m. is the circadian equivalent of asking you, their parent, to go to sleep at seven or eight p.m. No matter how loud you enunciate the order, no matter how much that teenager truly wishes to obey your instruction, and no matter what amount of willed effort is applied by either of the two parties, the circadian rhythm of a teenager will not be miraculously coaxed into a change. Furthermore, asking that same teenager to wake up at seven the next morning and function with intellect, grace, and good mood is the equivalent of asking you, their parent, to do the same at four or five a.m.

Sadly, neither society nor our parental attitudes are well designed to appreciate or accept that teenagers need more sleep than adults, and that they are biologically wired to obtain that sleep at a different time from their parents. It's very understandable for parents to feel frustrated in this way, since they believe that their teenager's sleep patterns reflect a conscious choice and not a biological edict. But non-volitional, non-negotiable, and strongly biological they are. We parents would be wise to accept this fact, and to embrace it, encourage it, and praise it, lest we wish our own children to suffer developmental brain abnormalities or force a raised risk of mental illness upon them.

It will not always be this way for the teenager. As they age into young and middle adulthood, their circadian schedule will gradually slide back in time. Not all the way back to the timing present in childhood, but back to an earlier schedule: one that, ironically, will lead those same (now) adults to have the same frustrations and annoyances with their own sons or daughters. By that stage, those parents have forgotten (or have chosen to forget) that they, too, were once adolescents who desired a much later bedtime than their own parents.

You may wonder why the adolescent brain first overshoots in their advancing circadian rhythm, staying up late and not wanting to wake up until late, yet will ultimately return to an earlier timed rhythm of sleep and wake in later adulthood. Though we continue to examine this question, the explanation I propose is a socio-evolutionary one.

Central to the goal of adolescent development is the transition from parental dependence to independence, all the while learning to navigate the complexities of peer-group relationships and interactions. One way in which Mother Nature has perhaps helped adolescents unbuckle themselves from their parents is to march their circadian rhythms forward in time, past that of their adult mothers and fathers. This ingenious biological solution selectively shifts teenagers to a later phase when they can, for several hours, operate independently—and do so as a peer-group collective. It is not a permanent or full dislocation from parental care, but as safe an attempt at partially separating soon-to-be adults from the eyes of Mother and Father. There is risk, of course. But the transition must happen. And the time of day when those independent adolescent wings unfold, and the first solo flights from the parental nest occur, is not a time of day at all, but rather a time of night, thanks to a forward-shifted circadian rhythm.

We are still learning more about the role of sleep in development. However, a strong case can already be made for defending sleep time in our adolescent youth, rather than denigrating sleep as a sign of laziness. As parents, we are often too focused on what sleep is taking away from our teenagers, without stopping to think about what it may be adding. Caffeine also comes into question. There was once an education policy in the US known as "No child left behind." Based on scientific evidence, a new policy has rightly been suggested by my colleague Dr. Mary Carskadon: "No child needs caffeine."

SLEEP IN MIDLIFE AND OLD AGE

As you, the reader, may painfully know; sleep is more problematic and disordered in older adults. The effects of certain medications more commonly taken by older adults, together with coexisting medical conditions, result in older adults being less able, on average, to obtain as much sleep, or as restorative a sleep, as young adults.

That older adults simply *need* less sleep is a myth. Older adults appear to need just as much sleep as they do in midlife, but are simply less able to generate that (still necessary) sleep. Affirming this, large surveys demonstrate that despite getting less sleep, older adults reported *needing*, and indeed *trying*, to obtain just as much sleep as younger adults.

There are additional scientific findings supporting the fact that older adults still need a full night of sleep, just like young adults, and I will address those shortly. Before I do, let me first explain the core impairments of sleep that occur with aging, and why those findings help falsify the argument that older adults don't need to sleep as much. These three key changes are: (1) reduced quantity/quality, (2) reduced sleep efficiency, and (3) disrupted timing of sleep.

The postadolescent stabilization of deep-NREM sleep in your early twenties does not remain very stable for very long. Soon—sooner than you may imagine or wish—comes a great sleep recession, with deep sleep being hit especially hard. In contrast to REM sleep, which remains largely stable in midlife, the decline of deep NREM sleep is already under way by your late twenties and early thirties.

As you enter your fourth decade of life, there is a palpable reduction in the electrical quantity and quality of that deep NREM sleep. You obtain fewer hours of deep sleep, and those deep NREM brainwaves become smaller, less powerful, and fewer in number. Passing into your mid- and late forties, age will have stripped you of 60 to 70 percent of the deep sleep you were enjoying as a young teenager. By the time you reach seventy years old, you will have lost 80 to 90 percent of your youthful deep sleep.

Certainly, when we sleep at night, and even when we wake in the morning, most of us do not have a good sense of our electrical sleep quality. Frequently this means that many seniors progress through their later years not fully realizing how degraded their deep-sleep quantity and quality have become. This is an important point: it means that elderly individuals fail to connect their deterioration in health with their deterioration in sleep, despite causal links between the two having been known to scientists for many decades. Seniors therefore complain about and seek treatment for their *health* issues when visiting their GP, but rarely ask for help with their equally problematic sleep issues. As a result, GPs are rarely motivated to address the problematic sleep in addition to the problematic health concerns of the older adult.

To be clear, not all medical problems of aging are attributable to poor sleep. But far more of our age-related physical and mental health ailments are related to sleep impairment than either we, or many doctors, truly realize or treat seriously. Once again, I urge an elderly individual who may be concerned about their sleep not to seek a sleeping pill prescription. Instead, I recommend you first explore the effective and scientifically proven non-pharmacological interventions that a doctor who is board certified in sleep medicine can provide.

The second hallmark of altered sleep as we age, and one that older adults are more conscious of, is *fragmentation*. The older we get, the more frequently we wake up throughout the night. There are many causes, including interacting medications and diseases, but chief among them is a weakened bladder. Older adults therefore visit the bathroom more frequently at night. Reducing fluid intake in the mid- and late evening can help, but it is not a cure-all.

Due to sleep fragmentation, older individuals will suffer a reduction in sleep efficiency, defined as the percent of time you were asleep while in bed. If you spent eight hours in bed, and slept for all eight of those hours, your sleep efficiency would be 100 percent. If you slept just four of those eight hours, your sleep efficiency would be 50 percent.

As healthy teenagers, we enjoyed a sleep efficiency of about 95 percent. As a reference anchor, most sleep doctors consider good-quality sleep to involve a sleep efficiency of 90 percent or above. By the time we reach our eighties, sleep efficiency has often dropped below 70 or 80 percent; 70 to 80 percent may sound reasonable until you realize that, within an eight-hour period in bed, it means you will spend as much as one to one and a half hours awake.

Inefficient sleep is no small thing, as studies assessing tens of thousands of older adults show. Even when controlling for factors such as body mass index, gender, race, history of smoking, frequency of exercise, and medications, the lower an older individual's sleep efficiency score, the higher their mortality risk, the worse their physical health, the more likely they are to suffer from depression, the less energy they report, and the lower their cognitive function, typified by forgetfulness.* Any individual, no matter what age, will exhibit physical ailments, mental health instability, reduced alertness, and impaired memory if their sleep is chronically disrupted. The problem in aging is that family members observe these daytime features in older relatives and jump to a diagnosis of dementia, overlooking the possibility that bad sleep is an equally likely cause. Not all old adults with sleep issues have dementia. But I will describe evidence in chapter 7 that clearly shows how and why sleep disruption is a causal factor contributing to dementia in mid- and later life.

A more immediate, though equally dangerous, consequence of fragmented sleep in the elderly warrants brief discussion: the nighttime bathroom visits and associated risk of falls and thus fractures. We are

---

*D. J. Foley, A. A. Monjan, S. L. Brown, E. M. Simonsick et al., "Sleep complaints among elderly persons: an epidemiologic study of three communities," *Sleep* 18, no. 6 (1995): 425–32. See also D. J. Foley, A. A. Monjan, E. M. Simonstick, R. B. Wallace, and D. G. Blazer, "Incidence and remission of insomnia among elderly adults: an epidemiologic study of 6,800 persons over three years," *Sleep* 22 (Suppl 2) (1999): S366–72.

often groggy when we wake up during the night. Add to this cognitive haze the fact that it is dark. Furthermore, having been recumbent in bed means that when you stand and start moving, blood can race from your head, encouraged by gravity, down toward your legs. You feel light-headed and unsteady on your feet as a consequence. The latter is especially true in older adults whose control of blood pressure is itself often impaired. All of these issues mean that an older individual is at a far higher risk of stumbling, falling, and breaking bones during nighttime visits to the bathroom. Falls and fractures markedly increase morbidity and significantly hasten the end of life of an older adult. In the footnotes, I offer a list of tips for safer nighttime sleep in the elderly.*

The third sleep change with advanced age is that of *circadian timing*. In sharp contrast to adolescents, seniors commonly experience a regression in sleep timing, leading to earlier and earlier bedtimes. The cause is an earlier evening release and peak of melatonin as we get older, instructing an earlier start time for sleep. Restaurants in retirement communities have long known of this age-related shift in bedtime preference, epitomized (and accommodated) by the "early-bird special."

Changes in circadian rhythms with advancing age may appear harmless, but they can be the cause of numerous sleep (and wake) problems in the elderly. Older adults often want to stay awake later into the evening so that they can go to theater or the movies, socialize, read, or watch television. But in doing so, they find themselves waking up on the couch, in a movie theater seat, or in a reclining chair, having inadvertently fallen asleep mid-evening. Their regressed circadian rhythm, instructed by an earlier release of melatonin, left them no choice.

But what seems like an innocent doze has a damaging consequence. The early-evening snooze will jettison precious sleep pressure, clearing

---

*Tips for safe sleep in the elderly: (1) have a side lamp within reach that you can switch on easily, (2) use dim or motion-activated night-lights in the bathrooms and hallways to illuminate your path, (3) remove obstacles or rugs en route to the bathroom to prevent stumbles or trips, and (4) keep a telephone by your bed with emergency phone numbers programmed on speed dial.

away the sleepiness power of adenosine that had been steadily building throughout the day. Several hours later, when that older individual gets into bed and tries to fall asleep, they may not have enough sleep pressure to fall asleep quickly, or stay asleep as easily. An erroneous conclusion follows: "I have insomnia." Instead, dozing off in the evening, which most older adults do not realize is classified as napping, can be the source of sleep difficulty, not true insomnia.

A compounding problem arrives in the morning. Despite having had trouble falling asleep that night and already running a sleep debt, the circadian rhythm—which, as you'll remember from chapter 2, operates independently of the sleep-pressure system—will start to rise around four or five a.m. in many elderly individuals, enacting its classic earlier schedule in seniors. Older adults are therefore prone to wake up early in the morning as the alerting drumbeat of the circadian rhythm grows louder, and corresponding hopes of returning back to sleep diminish in tandem.

Making matters worse, the strengths of the circadian rhythm and amount of nighttime melatonin released also decrease the older we get. Add these things up, and a self-perpetuating cycle ensues wherein many seniors are battling a sleep debt, trying to stay awake later in the evening, inadvertently dozing off earlier, finding it hard to fall or stay asleep at night, only to be woken up earlier than they wish because of a regressed circadian rhythm.

There are methods that can help push the circadian rhythm in older adults somewhat later, and also strengthen the rhythm. Here again, they are not a complete or perfect solution, I'm sad to say. Later chapters will describe the damaging influence of artificial light on the circadian twenty-four-hour rhythm (bright light at night). Evening light suppresses the normal rise in melatonin, pushing an average adult's sleep onset time into the early-morning hours, preventing sleep at a reasonable hour. However, this same sleep-delaying effect can be put to good use in older adults, if timed correctly. Having woken up early, many older adults are physically active during the morning hours, and therefore obtain much of their bright-light exposure in the first half of the day. This is not optimal, as it reinforces the early-to-rise, early-to-decline cycle of the twenty-four-hour internal clock. Instead, older adults who want to shift their bedtimes to a later hour should get bright-light exposure in the late-afternoon hours.

I am not, however, suggesting that older adults stop exercising in the morning. Exercise can help solidify good sleep, especially in the elderly. Instead, I advise two modifications for seniors. First, wear sunglasses during morning exercise outdoors. This will reduce the influence of morning light being sent to your suprachiasmatic clock that would otherwise keep you on an early-to-rise schedule. Second, go back outside in the late afternoon for sunlight exposure, but this time do not wear sunglasses. Make sure to wear sun protection of some sort, such as a hat, but leave the sunglasses at home. Plentiful later-afternoon daylight will help delay the evening release of melatonin, helping push the timing of sleep to a later hour.

Older adults may also wish to consult with their doctor about taking melatonin in the evening. Unlike young or middle-age adults, where melatonin has not proved efficacious for helping sleep beyond the circumstance of jet lag, prescription melatonin has been shown to help boost the otherwise blunted circadian and associated melatonin rhythm in the elderly, reducing the time taken to fall asleep and improving self-reported sleep quality and morning alertness.*

The change in circadian rhythm as we get older, together with more frequent trips to the bathroom, help to explain two of the three key nighttime issues in the elderly: early sleep onset/offset and sleep fragmentation. They do not, however, explain the first key change in sleep with advancing age: the loss of deep-sleep quantity and quality. Although scientists have known about the pernicious loss of deep sleep with advancing age for many decades, the cause has remained elusive: What is it about the aging process that so thoroughly robs the brain of this essential state of slumber? Beyond scientific curiosity, it is also a pressing clinical issue for the elderly, considering the importance of deep sleep for learning and memory, not to mention all branches of

---

*A. G. Wade, I. Ford, G. Crawford, et al., "Efficacy of prolonged release melatonin in insomnia patients aged 55–80 years: quality of sleep and next-day alertness outcomes," *Current Medical Research and Opinion* 23, no. 10: (2007): 2597–605.

bodily health, from cardiovascular and respiratory, to metabolic, energy balance, and immune function.

Working with an incredibly gifted team of young researchers, I set out to try and answer this question several years ago. I wondered whether the cause of this sleep decline was to be found in the intricate pattern of structural brain deterioration that occurs as we age. You will recall from chapter 3 that the powerful brainwaves of deep NREM sleep are generated in the middle-frontal regions of the brain, several inches above the bridge of your nose. We already knew that as individuals get older, their brains do not deteriorate uniformly. Instead, some parts of the brain start losing neurons much earlier and far faster than other parts of the brain—a process called atrophy. After performing hundreds of brain scans, and amassing almost a thousand hours of overnight sleep recordings, we discovered a clear answer, unfolding in a three-part story.

First, the areas of the brain that suffer the most dramatic deterioration with aging are, unfortunately, the very same deep-sleep-generating regions—the middle-frontal regions seated above the bridge of the nose. When we overlaid the map of brain degeneration hot spots in the elderly on the brain map that highlighted the deep-sleep-generating regions in young adults, there was a near-perfect match. Second, and unsurprisingly, older adults suffered a 70 percent loss of deep sleep, compared with matched young individuals. Third, and most critical, we discovered that these changes were not independent, but instead significantly connected with one another: the more severe the deterioration that an older adult suffers within this specific mid-frontal region of their brain, the more dramatic their loss of deep NREM sleep. It was a saddening confirmation of my theory: the parts of our brain that ignite healthy deep sleep at night are the very same areas that degenerate, or atrophy, earliest and most severely as we age.

In the years leading up to these investigations, my research team and several others around the world had demonstrated how critical deep sleep was for cementing new memories and retaining new facts in young adults. Knowing this, we had included a twist to our experiment in older adults. Several hours before going to sleep, all of these seniors learned a list of new facts (word associations), quickly followed by an immediate memory test to see how much information they had retained. The next morning, following the night of sleep recording, we tested them a second time. We could therefore determine the amount of memory savings that had occurred for any one individual across the night of sleep.

The older adults forgot far more of the facts by the following morning than the young adults—a difference of almost 50 percent. Furthermore, those older adults with the greatest loss of deep sleep showed the most catastrophic overnight forgetting. Poor memory and poor sleep in old age are therefore not coincidental, but rather significantly interrelated. The findings helped us shed new light on the forgetfulness that is all too common in the elderly, such as difficulty remembering people's names or memorizing upcoming hospital appointments.

It is important to note that the extent of brain deterioration in older adults explained 60 percent of their inability to generate deep sleep. This was a helpful finding. But the more important lesson to be gleaned from this discovery for me was that 40 percent of the explanation for the loss of deep sleep in the elderly remained unaccounted for by our discovery. We are now hard at work trying to discover what that is. Recently, we identified one factor—a sticky, toxic protein that builds up in the brain called beta-amyloid that is a key cause of Alzheimer's disease: a discovery discussed in the next several chapters.

More generally, these and similar studies have confirmed that poor sleep is one of the most underappreciated factors contributing to cognitive and medical ill health in the elderly, including issues of diabetes, depression, chronic pain, stroke, cardiovascular disease, and Alzheimer's disease.

An urgent need therefore exists for us to develop new methods that restore some quality of deep, stable sleep in the elderly. One promising example that we have been developing involves brain stimulation methods, including controlled electrical stimulation pulsed into the brain at night. Like a supporting choir to a flagging lead vocalist, our goal is to electrically sing (stimulate) in time with the ailing brainwaves of older adults, amplifying the quality of their deep brainwaves and salvaging the health- and memory-promoting benefits of sleep.

Our early results look cautiously promising, though much, much more work is required. With replication, our findings can further debunk the long-held belief that we touched on earlier: older adults need less sleep. This myth has stemmed from certain observations that, to some scientists, suggest that an eighty-year-old, say, simply needs less sleep than a fifty-year-old. Their arguments are as follows. First, if you deprive older adults of sleep, they do not show as dramatic an impairment in performance on a basic response-time task as a younger adult. Therefore, older adults must need sleep less than younger adults. Second, older adults generate less sleep than young adults, so by inference, older adults must simply need sleep less. Third, older adults do not show as strong a sleep rebound after a night of deprivation compared with young adults. The conclusion was that seniors therefore have less need for sleep if they have less of a recovery rebound.

There are, however, alternative explanations. Using performance as a measure of sleep need is perilous in older adults, since older adults are already impaired in their reaction times to begin with. Said unkindly, older adults don't have much further to fall in terms of getting worse, sometimes called a "floor effect," making it difficult to estimate the real performance impact of sleep deprivation.

Next, just because an older individual obtains less sleep, or does not obtain as much recovery sleep after sleep deprivation, does not necessarily mean that their *need* for sleep is less. It may just as easily indicate that they cannot physiologically *generate* the sleep they still nevertheless need. Take the alternative example of bone density, which is lower in older compared with younger adults. We do not assume that older individuals need weaker bones just because they have reduced bone density. Nor do we believe that older adults have bones that are weaker simply because they don't recover bone density and heal as quickly as young adults after suffering a fracture or break. Instead, we realize that their bones, like the centers of the brain that produce sleep, deteriorate with age, and we accept this degeneration as the cause of numerous health issues. We consequently provide dietary supplements, physical therapy, and medications to try to offset bone deficiency. I believe we should recognize and treat sleep impairments in the elderly with a similar regard and compassion, recognizing that they do, in fact, need just as much sleep as other adults.

Finally, the preliminary results of our brain stimulation studies suggest that older adults, may, in fact, need more sleep than they themselves can naturally generate, since they benefit from an improvement in sleep quality, albeit through artificial means. If older individuals did not need more deep sleep, then they should already be satiated, and not benefit from receiving more (artificially, in this case). Yet they do benefit from having their sleep enhanced, or perhaps worded correctly, restored. That is, older adults, and especially those with different forms of dementia, appear to suffer an unmet sleep need, which demands new treatment options: a topic that we shall soon return to.

PART 2

Why Should You Sleep?

CHAPTER 6

Your Mother and Shakespeare Knew

The Benefits of Sleep for the Brain

AMAZING BREAKTHROUGH!

Scientists have discovered a revolutionary new treatment that makes you live longer. It enhances your memory and makes you more creative. It makes you look more attractive. It keeps you slim and lowers food cravings. It protects you from cancer and dementia. It wards off colds and the flu. It lowers your risk of heart attacks and stroke, not to mention diabetes. You'll even feel happier, less depressed, and less anxious. Are you interested?

While it may sound hyperbolic, nothing about this fictitious advertisement would be inaccurate. If it were for a new drug, many people would be disbelieving. Those who were convinced would pay large sums of money for even the smallest dose. Should clinical trials back up the claims, share prices of the pharmaceutical company that invented the drug would skyrocket.

Of course, the ad is not describing some miracle new tincture or a cure-all wonder drug, but rather the proven benefits of a full night of sleep. The evidence supporting these claims has been documented in more than 17,000 well-scrutinized scientific reports to date. As for the prescription cost, well, there isn't one. It's free. Yet all too often, we shun the nightly invitation to receive our full dose of this all-natural remedy—with terrible consequences.

Failed by the lack of public education, most of us do not realize how remarkable a panacea sleep truly is. The following three chapters are designed to help rectify our ignorance born of this largely absent public health message. We will come to learn that sleep is *the* universal health care provider: whatever the physical or mental ailment, sleep has a prescription it can dispense. Upon completion of these chapters, I hope even the most ardent of short-sleepers will be swayed, having a reformed deference.

Earlier, I described the component stages of sleep. Here, I reveal the attendant virtues of each. Ironically, most all of the "new," twenty-first-century discoveries regarding sleep were delightfully summarized in 1611 in *Macbeth*, act two, scene two, where Shakespeare prophetically states that sleep is "the chief nourisher in life's feast."* Perhaps, with less highfalutin language, your mother offered similar advice, extolling the benefits of sleep in healing emotional wounds, helping you learn and remember, gifting you with solutions to challenging problems, and preventing sickness and infection. Science, it seems, has simply been evidential, providing proof of everything your mother, and apparently Shakespeare, knew about the wonders of sleep.

SLEEP FOR THE BRAIN

Sleep is not the absence of wakefulness. It is far more than that. Described earlier, our nighttime sleep is an exquisitely complex, metabolically active, and deliberately ordered series of unique stages.

Numerous functions of the brain are restored by, and depend upon, sleep. No one type of sleep accomplishes all. Each stage of sleep—light NREM sleep, deep NREM sleep, and REM sleep—offer different brain benefits at different times of night. Thus, no one type of sleep is more essential than another. Losing out on any one of these types of sleep will cause brain impairment.

Of the many advantages conferred by sleep on the brain, that of memory is especially impressive, and particularly well understood. Sleep has proven itself time and again as a memory aid: both before

*"Sleep that knits up the ravell'd sleeve of care,
The death of each day's life, sore labour's bath,
Balm of hurt minds, great nature's second course,
Chief nourisher in life's feast,—"
William Shakespeare, *Macbeth*, Folger Shakespeare Library (New York: Simon & Schuster; first edition, 2003).

learning, to prepare your brain for initially making new memories, and after learning, to cement those memories and prevent forgetting.

SLEEP-THE-NIGHT-BEFORE LEARNING

Sleep *before* learning refreshes our ability to initially make new memories. It does so each and every night. While we are awake, the brain is constantly acquiring and absorbing novel information (intentionally or otherwise). Passing memory opportunities are captured by specific parts of the brain. For fact-based information—or what most of us think of as textbook-type learning, such as memorizing someone's name, a new phone number, or where you parked your car—a region of the brain called the hippocampus helps apprehend these passing experiences and binds their details together. A long, finger-shaped structure tucked deep on either side of your brain, the hippocampus offers a short-term reservoir, or temporary information store, for accumulating new memories. Unfortunately, the hippocampus has a limited storage capacity, almost like a camera roll or, to use a more modern-day analogy, a USB memory stick. Exceed its capacity and you run the risk of not being able to add more information or, equally bad, overwriting one memory with another: a mishap called interference forgetting.

How, then, does the brain deal with this memory capacity challenge? Some years ago, my research team wondered if sleep helped solve this storage problem by way of a file-transfer mechanism. We examined whether sleep shifted recently acquired memories to a more permanent, long-term storage location in the brain, thereby freeing up our short-term memory stores so that we awake with a refreshed ability for new learning.

We began testing this theory using daytime naps. We recruited a group of healthy young adults and randomly divided them into a nap group and a no-nap group. At noon, all the participants underwent a rigorous session of learning (one hundred face-name pairs) intended to tax the hippocampus, their short-term memory storage site. As expected, both groups performed at comparable levels. Soon after, the nap group took a ninety-minute siesta in the sleep laboratory with electrodes placed on their heads to measure sleep. The no-nap group stayed awake in the laboratory and performed menial activities, such as browsing the Internet or playing board games. Later that day, at six p.m., all participants performed another round of intensive learning where they tried to cram yet another set of new facts into their short-term storage reservoirs (another one hundred face-name pairs). Our question was simple: Does the learning capacity of the human brain decline with continued time awake across the day and, if so, can sleep reverse this saturation effect and thus restore learning ability?

Those who were awake throughout the day became progressively worse at learning, even though their ability to concentrate remained stable (determined by separate attention and response time tests). In contrast, those who napped did markedly better, and actually improved in their capacity to memorize facts. The difference between the two groups at six p.m. was not small: a 20 percent learning advantage for those who slept.

Having observed that sleep restores the brain's capacity for learning, making room for new memories, we went in search of exactly what it was about sleep that transacted the restoration benefit. Analyzing the electrical brainwaves of those in the nap group brought our answer. The memory refreshment was related to lighter, stage 2 NREM sleep, and specifically the short, powerful bursts of electrical activity called sleep spindles, noted in chapter 3. The more sleep spindles an individual obtained during the nap, the greater the restoration of their learning when they woke up. Importantly, sleep spindles did not predict someone's innate learning aptitude. That would be a less interesting result, as it would imply that inherent learning ability and spindles simply go hand in hand. Instead, it was specifically the *change* in learning from before relative to after sleep, which is to say the *replenishment* of learning ability, that spindles predicted.

Perhaps more remarkable, as we analyzed the sleep-spindle bursts of activity, we observed a strikingly reliable loop of electrical current pulsing throughout the brain that repeated every 100 to 200 milliseconds. The pulses kept weaving a path back and forth between the hippocampus, with its short-term, limited storage space, and the far larger, long-term storage site of the cortex (analogous to a large-memory hard drive).* In that moment, we had just become privy to an electrical transaction occurring in the quiet secrecy of sleep: one that was shifting fact-based memories from the temporary storage depot (the hippocampus) to a long-term secure vault (the cortex). In doing so, sleep had delightfully cleared out the hippocampus, replenishing this short-term information repository with plentiful free space. Participants awoke with a refreshed capacity to absorb new information within the hippocampus, having relocated yesterday's imprinted experiences to a more permanent safe hold. The learning of new facts could begin again, anew, the following day.

We and other research groups have since repeated this study across a full night of sleep and replicated the same finding: the more sleep spindles an individual has at night, the greater the restoration of overnight learning ability come the next morning.

Our recent work on this topic has returned to the question of aging. We have found that seniors (aged sixty to eighty years old) are unable to generate sleep spindles to the same degree as young, healthy adults, suffering a 40 percent deficit. This led to a prediction: the fewer sleep spindles an older adult has on a particular night, the harder it should be for them to cram new facts into their hippocampus the next day, since they have not received as much overnight refreshment of short-term memory capacity. We conducted the study, and that is precisely what we found: the fewer the number of spindles an elderly brain produced on a particular night, the lower the learning capacity of that older individual the next day, making it more difficult for them to memorize the list of facts we presented. This sleep and learning link is yet one more reason for medicine to take more seriously the sleep complaints of the elderly, further compelling researchers such as myself to find new, non-pharmacological methods for improving sleep in aging populations worldwide.

Of broader societal relevance, the concentration of NREM-sleep spin-

*The literal-minded reader should not take this analogy to suggest that I believe the human brain, or even its functions of learning and memory, operates as a computer does. There are abstract similarities, yes, but there are many clear differences, large and small. A brain cannot be said to be the equivalent of a computer, nor vice versa. It is simply that certain conceptual parallels offer useful analogies to comprehend how the biological processes of sleep operate.

dles is especially rich in the late-morning hours, sandwiched between long periods of REM sleep. Sleep six hours or less and you are short-changing the brain of a learning restoration benefit that is normally performed by sleep spindles. I will return to the broader educational ramifications of these findings in a later chapter, addressing the question of whether early school start times, which throttle precisely this spindle-rich phase of sleep, are optimal for the teaching of young minds.

SLEEP-THE-NIGHT-AFTER LEARNING

The second benefit of sleep for memory comes *after* learning, one that effectively clicks the "save" button on those newly created files. In doing so, sleep protects newly acquired information, affording immunity against forgetting: an operation called consolidation. That sleep sets in motion the process of memory consolidation was recognized long ago, and may be one of the oldest proposed functions of sleep. The first such claim in the written human record appears to be by the prophetic Roman rhetorician Quintilian (AD 35–100), who stated:

> It is a curious fact, of which the reason is not obvious, that the interval of a single night will greatly increase the strength of the memory. . . . Whatever the cause, things which could not be recalled on the spot are easily coordinated the next day, and time itself, which is generally accounted one of the causes of forgetfulness, actually serves to strengthen the memory.*

It was not until 1924 when two German researchers, John Jenkins and Karl Dallenbach, pitted sleep and wake against each other to see which one won out for a memory-savings benefit—a memory researchers' version of the classic Coke vs. Pepsi challenge. Their study participants first learned a list of verbal facts. Thereafter, the researchers tracked how quickly the participants forgot those memories over an eight-hour time interval, either spent awake or across a night of sleep. Time spent

*Nicholas Hammond, *Fragmentary Voices: Memory and Education at Port-Royal* (Tübingen, Germany: Narr Dr. Gunter; 2004).

asleep helped cement the newly learned chunks of information, preventing them from fading away. In contrast, an equivalent time spent awake was deeply hazardous to recently acquired memories, resulting in an accelerated trajectory of forgetting.*

The experimental results of Jenkins and Dallenbach have now been replicated time and again, with a memory retention benefit of between 20 and 40 percent being offered by sleep, compared to the same amount of time awake. This is not a trivial concept when you consider the potential advantages in the context of studying for an exam, for instance, or evolutionarily, in remembering survival-relevant information such as the sources of food and water and locations of mates and predators.

It was not until the 1950s, with the discovery of NREM and REM sleep, that we began to understand more about *how*, rather than simply *if*, sleep helps to solidify new memories. Initial efforts focused on deciphering what stage(s) of sleep made immemorial that which we had imprinted onto the brain during the day, be it facts in the classroom, medical knowledge in a residency training program, or a business plan from a seminar.

You will recall from chapter 3 that we obtain most of our deep NREM sleep early in the night, and much of our REM sleep (and lighter NREM sleep) late in the night. After having learned lists of facts, researchers allowed participants the opportunity to sleep only for the first half of the night or only for the second half of the night. In this way, both experimental groups obtained the same total amount of sleep (albeit short), yet the former group's sleep was rich in deep NREM, and the latter was dominated instead by REM. The stage was set for a battle royal between the two types of sleep. The question: Which sleep period would confer a greater memory savings benefit—that filled with deep NREM, or that packed with abundant REM sleep? For fact-based, textbook-like memory, the result was clear. It was early-night sleep, rich in deep NREM, that won out in terms of providing superior memory retention savings relative to late-night, REM-rich sleep.

Investigations in the early 2000s arrived at a similar conclusion using

*J. G. Jenkins and K. M. Dallenbach, "Obliviscence during sleep and waking," *American Journal of Psychology* 35 (1924): 605–12.

a slightly different approach. Having learned a list of facts before bed, participants were allowed to sleep a full eight hours, recorded with electrodes placed on the head. The next morning, participants performed a memory test. When researchers correlated the intervening sleep stages with the number of facts retained the following morning, deep NREM sleep carried the vote: the more deep NREM sleep, the more information an individual remembered the next day. Indeed, if you were a participant in such a study, and the only information I had was the amount of deep NREM sleep you had obtained that night, I could predict with high accuracy how much you would remember in the upcoming memory test upon awakening, even before you took it. That's how deterministic the link between sleep and memory consolidation can be.

Using MRI scans, we have since looked deep into the brains of participants to see where those memories are being retrieved from before sleep relative to after sleep. It turns out that those information packets were being recalled from very different geographical locations within the brain at the two different times. Before having slept, participants were fetching memories from the short-term storage site of the hippocampus—that temporary warehouse, which is a vulnerable place to live for any long duration of time if you are a new memory. But things looked very different by the next morning. The memories had moved. After the full night of sleep, participants were now retrieving that same information from the neocortex, which sits at the top of the brain—a region that serves as the long-term storage site for fact-based memories, where they can now live safely, perhaps in perpetuity.

We had observed a real-estate transaction that takes place each night when we sleep. Fitting the notion of a long-wave radio signal that carries information across large geographical distances, the slow brainwaves of deep NREM had served as a courier service, transporting memory packets from a temporary storage hold (hippocampus) to a more secure, permanent home (the cortex). In doing so, sleep had helped future-proof those memories.

Put these findings together with those I described earlier regarding initial memorization, and you realize that the anatomical dialogue established during NREM sleep (using sleep spindles and slow waves) between the hippocampus and cortex is elegantly synergistic. By transferring memories of yesterday from the short-term repository of the hippocampus to the long-term home within the cortex, you awake with both yesterday's experiences safely filed away and having regained your short-term storage capacity for new learning throughout that following day. The cycle repeats each day and night, clearing out the cache of short-term memory for the new imprinting of facts, while accumulating an ever-updated catalog of past memories. Sleep is constantly modifying the information architecture of the brain at night. Even daytime naps as short as twenty minutes can offer a memory consolidation advantage, so long as they contain enough NREM sleep.*

Study infants, young kids, or adolescents and you see the very same overnight memory benefit of NREM sleep, sometimes even more powerfully so. For those in midlife, forty- to sixty-year-olds, deep NREM sleep continues to help the brain retain new information in this way, with the decline in deep NREM sleep and the deterioration in the ability to learn and retain memories in old age having already been discussed.

At every stage of human life, the relationship between NREM sleep and memory solidification is therefore observed. It's not just humans, either. Studies in chimpanzees, bonobos, and orangutans have demonstrated that all three groups are better able to remember where food items have been placed in their environments by experimenters after they sleep.† Descend down the phylogenetic chain to cats, rats, and even insects, and the memory-maintaining benefit of NREM sleep remains on powerful display.

Though I still marvel at Quintilian's foresight and straightforward description of what scientists would, thousands of years later, prove true about sleep's benefit to memory, I prefer the words of two equally accomplished philosophers of their time, Paul Simon and Art Garfunkel. In February of 1964, they penned a now famous set of lyrics that encapsulate the same nocturnal event in the song "The Sound of Silence." Perhaps you know the song and lyrics. Simon and Garfunkel describe greeting their old friend, darkness (sleep). They speak of relaying the

---

*Such findings may offer cognitive justification for the common incidence of unintentional napping in public in Japanese culture, termed *inemuri* ("sleep while being present").

†G. Martin-Ordas and J. Call, "Memory processing in great apes: the effect of time and sleep," *Biology Letters* 7, no. 6 (2011): 829–32.

day's waking events to the sleeping brain at night in the form of a vision, softly creeping—a gentle information upload, if you will. Insightfully, they illustrate how those fragile seeds of waking experience, sown during the day, have now been embedded ("planted") in the brain during sleep. As a result of that process, those experiences now remain upon awakening the next morning. Sleep's future-proofing of memories, all packaged for us in perfect song lyrics.

A slight, but important, modification to Simon and Garfunkel's lyrics is warranted, based on very recent evidence. Not only does sleep *maintain* those memories you have successfully learned before bed (*"the vision that was planted in my brain / Still remains")*, but it will even salvage those that appeared to have been lost soon after learning. In other words, following a night of sleep you regain access to memories that you could not retrieve before sleep. Like a computer hard drive where some files have become corrupted and inaccessible, sleep offers a recovery service at night. Having repaired those memory items, rescuing them from the clutches of forgetting, you awake the next morning able to locate and retrieve those once unavailable memory files with ease and precision. The "ah yes, now I remember" sensation that you may have experienced after a good night of sleep.

Having narrowed in on the type of sleep—NREM sleep—responsible for making fact-based memories permanent, and further recovering those that were in jeopardy of being lost, we have begun exploring ways to experimentally boost the memory benefits of sleep. Success has come in two forms: sleep stimulation, and targeted memory reactivation. The clinical ramifications of both will become clear when considered in the context of psychiatric illness and neurological disorders, including dementia.

Since sleep is expressed in patterns of electrical brainwave activity, sleep stimulation approaches began by trading in the same currency: electricity. In 2006, a research team in Germany recruited a group of healthy young adults for a pioneering study in which they applied electrode pads onto the head, front and back. Rather than recording the electrical brainwaves being emitted from the brain during sleep, the scientists did the opposite: inserted small amounts of electrical voltage. They patiently waited until each participant had entered into the deepest stages of NREM sleep and, at that point, switched on the brain stimulator, pulsing in rhythmic time with the slow waves. The electrical pulsations were so small that participants did not feel them, nor did they wake up.* But they had a measurable impact on sleep.

Both the size of the slow brainwaves and the number of sleep spindles riding on top of the deep brainwaves were increased by the stimulation, relative to a control group of subjects who did not receive stimulation during sleep. Before being put to bed, all the participants had learned a list of new facts. They were tested the next morning after sleep. By boosting the electrical quality of deep-sleep brainwave activity, the researchers almost doubled the number of facts that individuals were able to recall the following day, relative to those participants who received no stimulation. Applying stimulation during REM sleep, or during wakefulness across the day, did not offer similar memory advantages. Only stimulation during NREM sleep, in synchronous time with the brain's own slow mantra rhythm, leveraged a memory improvement.

Other methods for amplifying the brainwaves of sleep are fast being developed. One technology involves quiet auditory tones being played over speakers next to the sleeper. Like a metronome in rhythmic stride with the individual slow waves, the tick-tock tones are precisely synchronized with the individual's sleeping brainwaves to help entrain their rhythm and produce even deeper sleep. Relative to a control group that slept but had no synchronous auditory chimes at night, the auditory stimulation increased the power of the slow brainwaves and returned an impressive 40 percent memory enhancement the next morning.

Before you drop this book and start installing speakers above your bed, or go shopping for an electrical brain stimulator, let me dissuade you. For both methods, the wisdom of "do not try this at home" applies. Some individuals have made their own brain-stimulating devices, or bought such devices online, which are not covered by safety regulations. Skin burns and temporary losses of vision have been reported by

---

*This technique, called transcranial direct current brain stimulation (tDCS), should not be confused with electroconvulsive shock therapy, in which the size of electrical voltage inserted into the brain is many hundreds of times stronger (a depiction of which is illustrated in Jack Nicholson's performance in the movie *One Flew Over the Cuckoo's Nest*).

mistakes in construction or voltage application. Playing loud tick-tock acoustic tones on repeat next to your bed sounds like a safer option, but you may be doing more harm than good. When researchers in the above studies timed the auditory tones to strike just off the natural peak of each slow brainwave, rather than in perfect time with each brainwave, they disrupted, rather than enhanced, sleep quality.

If brain stimulation or auditory tones were not bizarre enough, a Swiss research team recently suspended a bedframe on ropes from the ceiling of a sleep laboratory (stick with me here). Affixed to one side of the suspended bed was a rotating pulley. It allowed the researchers to sway the bed from side to side at controlled speeds. Volunteers then took a nap in the bed as the researchers recorded their sleeping brainwaves. In half of the participants, the researchers gently rocked the bed once they entered NREM sleep. In the other half of the subjects, the bed remained static, offering a control condition. Slow rocking increased the depth of deep sleep, boosted the quality of slow brainwaves, and more than doubled the number of sleep spindles. It is not yet known whether these sway-induced sleep changes enhance memory, since the researchers did not perform any such tests with their participants. Nevertheless, the findings offer a scientific explanation for the ancient practice of rocking a child back and forth in one's arms, or in a crib, inducing a deep sleep.

Sleep stimulation methods are promising, but they do have a potential limitation: the memory benefit they provide is indiscriminate. That is, all things learned before sleep are generally enhanced the next day. Similar to a prix fixe menu at a restaurant in which there are no options, you are going to get served all dishes listed, like it or not. Most people do not enjoy this type of food service, which is why most restaurants offer you a large menu from which you can pick and choose, selecting only those items you would like to receive.

What if a similar opportunity was possible with sleep and memory? Before going to bed, you would review the learning experiences of the day, selecting only those memories from the menu list that you would like improved. You place your order, then go to sleep, knowing that your order will be served to you overnight. When you wake up in the morning, your brain will have been nourished only by the specific items you ordered from the autobiographical carte du jour. You have, as a consequence, selectively enhanced only those individual memories that you want to keep. It all sounds like the stuff of science fiction, but it is now science fact: the method is called targeted memory reactivation. And as is so often the case, the true story turns out to be far more fascinating than the fictional one.

Before going to sleep, we show participants individual pictures of objects at different spatial locations on a computer screen, such as a cat in the lower right side, or a bell in the upper center, or a kettle near the top right of the screen. As a participant, you have to remember not only the individual items you have been shown, but also their spatial location on the screen. You will be shown a hundred of these items. After sleep, picture objects will again appear on the screen, now in the center, some of which you have seen before, some you have not. You have to decide if you remember the picture object or not, and if you do, you must move that picture object to the spatial location on the screen where it originally appeared, using a mouse. In this way, we can assess whether you remember the object, and also how accurately you can remember its location.

But here is the intriguing twist. As you were originally learning the images before sleep, each time an object was presented on the screen, a corresponding sound was played. For example, you would hear "meow" when the cat picture was shown, or "ding-a-ling" when the bell was shown. All picture objects are paired, or "auditory-tagged," with a semantically matching sound. When you are asleep, and in NREM sleep specifically, an experimenter will replay half of the previously tagged sounds (fifty of the total hundred) to your sleeping brain at low volume using speakers on either side of the bed. As if helping guide the brain in a targeted search-and-retrieve effort, we can trigger the selective reactivation of corresponding individual memories, prioritizing them for sleep-strengthening, relative to those that were not reactivated during NREM sleep.

When you are tested the following morning, you will have a quite remarkable bias in your recollection, remembering far more of the items that we reactivated during sleep using the sound cues than those not reactivated. Note that all one hundred of the original memory items passed through sleep. However, using sound cuing, we avoid indiscriminate enhancement of all that you learned. Analogous to looping your favorite songs in a repeating playlist at night, we cherry-pick specific slices of your autobiographical past, and preferentially strengthen them by using the individualized sound cues during sleep.*

I'm sure you can imagine innumerable uses for such a method. That said, you may also feel ethically uncomfortable about the prospect, considering that you would have the power to write and rewrite your own remembered life narrative or, more concerning, that of someone else. This moral dilemma is somewhat far in the future, but should such methods continue to be refined, it is one we may face.

SLEEP TO FORGET?

Up to this point, we have discussed the power of sleep after learning to enhance remembering and avoid forgetting. However, the capacity to forget can, in certain contexts, be as important as the need for remembering, both in day-to-day life (e.g., forgetting last week's parking spot in preference for today's) and clinically (e.g., in excising painful, disabling memories, or in extinguishing craving in addiction disorders). Moreover, forgetting is not just beneficial to delete stored information we no longer need. It also lowers the brain resources required for retrieving those memories we want to retain, similar to the ease of finding important documents on a neatly organized, clutter-free desk. In this way, sleep helps you retain everything you need and nothing that you don't, improving the ease of memory recollection. Said another way, forgetting is the price we pay for remembering.

In 1983, the Nobel Laureate Francis Crick, who discovered the helical structure of DNA, decided to turn his theoretical mind toward the topic of sleep. He suggested that the function of REM-sleep dreaming was to remove unwanted or overlapping copies of information in the brain: what he termed "parasitic memories." It was a fascinating idea, but it remained just that—an idea—for almost thirty years, receiving

*This nighttime reactivation method only works during NREM sleep and does not work if attempted during REM sleep.

no formal examination. In 2009, a young graduate student and I put the hypothesis to the test. The results brought more than a few surprises.

We designed an experiment that again used daytime naps. At midday, our research subjects studied a long list of words presented one at a time on a computer screen. After each word had been presented on the screen, however, a large green "R" or a large red "F" was displayed, indicating to the participant that they should remember the prior word (R) or forget the prior word (F). It is not dissimilar to being in a class and, after having been told a fact, the teacher impresses upon you that it is especially important to remember that information for the exam, or instead that they made an error and the fact was incorrect, or the fact will not be tested on the exam, so you don't need to worry about remembering it for the test. We were effectively doing the same thing for each word right after learning, tagging it with the label "to be remembered" or "to be forgotten."

Half of the participants were then allowed a ninety-minute afternoon nap, while the other half remained awake. At six p.m. we tested everyone's memory for all of the words. We told participants that regardless of the tag previously associated with a word—to be remembered or to be forgotten—they should try to recall as many words as possible. Our question was this: Does sleep improve the retention of all words equally, or does sleep obey the waking command only to remember some items while forgetting others, based on the tags we had connected to each?

The results were clear. Sleep powerfully, yet very selectively, boosted the retention of those words previously tagged for "remembering," yet actively avoided the strengthening of those memories tagged for "forgetting." Participants who did not sleep showed no such impressive parsing and differential saving of the memories.*

We had learned a subtle, but important, lesson: sleep was far more intelligent than we had once imagined. Counter to earlier assumptions in the twentieth and twenty-first centuries, sleep does not offer a general, nonspecific (and hence verbose) preservation of all the information you learn during the day. Instead, sleep is able to offer a far more discerning hand in memory improvement: one that preferentially picks and chooses what

*You can even pay participants for each word they correctly recall to try and override what may be a simple reporting bias, and the results don't change.

information is, and is not, ultimately strengthened. Sleep accomplishes this by using meaningful tags that have been hung onto those memories during initial learning, or potentially identified during sleep itself. Numerous studies have shown a similarly intelligent form of sleep-dependent memory selection across both daytime naps and a full night of sleep.

When we analyzed the sleep records of those individuals who napped, we gained another insight. Contrary to Francis Crick's prediction, it was not REM sleep that was sifting through the list of prior words, separating out those that should be retained and those that should be removed. Rather, it was NREM sleep, and especially the very quickest of the sleep spindles that helped bend apart the curves of remembering and forgetting. The more of those spindles a participant had during a nap, the greater the efficiency with which they strengthened items tagged for remembering and actively eliminated those designated for forgetting.

Exactly how sleep spindles accomplish this clever memory trick remains unclear. What we have at least discovered is a rather telling pattern of looping activity in the brain that coincides with these speedy sleep spindles. The activity circles between the memory storage site (the hippocampus) and those regions that program the decision of intentionality (in the frontal lobe), such as "This is important" or "This is irrelevant." The recursive cycle of activity between these two areas (memory and intentionality), which happens ten to fifteen times per second during the spindles, may help explain NREM sleep's discerning memory influence. Much like selecting intentional filters on an Internet search or a shopping app, spindles offer a refining benefit to memory by allowing the storage site of your hippocampus to check in with the intentional filters carried in your astute frontal lobes, allowing selection only of that which you need to save, while discarding that which you do not.

We are now exploring ways of harnessing this remarkably intelligent service of selective remembering and forgetting with painful or problematic memories. The idea may invoke the premise of the Oscar-winning movie *Eternal Sunshine of the Spotless Mind*, in which individuals can have unwanted memories deleted by a special brain-scanning machine. In contrast, my real-world hope is to develop accurate methods for selectively weakening or erasing certain memories from an individual's memory library when there is a confirmed clinical need, such as in trauma, drug addiction, or substance abuse.

SLEEP FOR OTHER TYPES OF MEMORY

All of the studies I have described so far deal with one type of memory—that for facts, which we associate with textbooks or remembering someone's name. There are, however, many other types of memory within the brain, including skill memory. Take riding a bike, for example. As a child, your parents did not give you a textbook called *How to Ride a Bike*, ask you to study it, and then expect you to immediately begin riding your bike with skilled aplomb. Nobody can tell you how to ride a bike. Well, they can try, but it will do them—and more importantly you—no good. You can only learn how to ride a bike by doing rather than reading. Which is to say by practicing. The same is true for all motor skills, whether you are learning a musical instrument, an athletic sport, a surgical procedure, or how to fly a plane.

The term "muscle memory" is a misnomer. Muscles themselves have no such memory: a muscle that is not connected to a brain cannot perform any skilled actions, nor does a muscle store skilled routines. Muscle memory is, in fact, brain memory. Training and strengthening muscles can help you better *execute* a skilled memory routine. But the *routine* itself—the memory program—resides firmly and exclusively within the brain.

Years before I explored the effects of sleep on fact-based, textbook-like learning, I examined motor skill memory. Two experiences shaped my decision to perform these studies. The first was given to me as a young student at the Queen's Medical Center—a large teaching hospital in Nottingham, England. Here, I performed research on the topic of movement disorders, specifically spinal-cord injury. I was trying to discover ways of reconnecting spinal cords that had been severed, with the ultimate goal of reuniting the brain with the body. Sadly, my research was a failure. But during that time, I learned about patients with varied forms of motor disorders, including stroke. What struck me about so many of these patients was an iterative, step-by-step recovery of their motor function after the stroke, be it legs, arms, fingers, or speech.

Rarely was the recovery complete, but day by day, month by month, they all showed some improvement.

The second transformative experience happened some years later while I was obtaining my PhD. It was 2000, and the scientific community had proclaimed that the next ten years would be "The Decade of the Brain," forecasting (accurately, as it turned out) what would be remarkable progress within the neurosciences. I had been asked to give a public lecture on the topic of sleep at a celebratory event. At the time, we still knew relatively little about the effects of sleep on memory, though I made brief mention of the embryonic findings that were available.

After my lecture, a distinguished-looking gentleman with a kindly affect, dressed in a tweed suit jacket with a subtle yellow-green hue that I still vividly recall to this day, approached me. It was a brief conversation, but one of the most scientifically important of my life. He thanked me for the presentation, and told me that he was a pianist. He said he was intrigued by my description of sleep as an active brain state, one in which we may review and even strengthen those things we have previously learned. Then came a comment that would leave me reeling, and trigger a major focus of my research for years to come. "As a pianist," he said, "I have an experience that seems far too frequent to be chance. I will be practicing a particular piece, even late into the evening, and I cannot seem to master it. Often, I make the same mistake at the same place in a particular movement. I go to bed frustrated. But when I wake up the next morning and sit back down at the piano, I can just play, perfectly."

"I can just play." The words reverberated in my mind as I tried to compose a response. I told the gentleman that it was a fascinating idea, and it was certainly possible that sleep assisted musicianship and led to error-free performance, but that I knew of no scientific evidence to support the claim. He smiled, seeming unfazed by the absence of empirical affirmation, thanked me again for my lecture, and walked toward the reception hall. I, on the other hand, remained in the auditorium, realizing that this gentleman had just told me something that violated the most repeated and entrusted teaching edict: practice makes perfect. Not so, it seemed. Perhaps it was practice, *with sleep*, that makes perfect?

After three years of subsequent research, I published a paper with a similar title, and in the studies that followed gathered evidence that ultimately confirmed all of the pianist's wonderful intuitions about sleep. The findings also shed light on how the brain, after injury or damage by a stroke, gradually regains some ability to guide skill movements day by day—or should I say, night by night.

By that time, I had taken a position at Harvard Medical School, and with Robert Stickgold, a mentor and now a longtime collaborator and friend, we set about trying to determine if and how the brain continues to learn in the absence of any further practice. Time was clearly doing something. But it seemed that there were, in fact, three distinct possibilities to discriminate among. Was it (1) time, (2) time awake, or (3) time asleep that incubated skilled memory perfection?

I took a large group of right-handed individuals and had them learn to type a number sequence on a keyboard with their left hand, such as 4-1-3-2-4, as quickly and as accurately as possible. Like learning a piano scale, subjects practiced the motor skill sequence over and over again, for a total of twelve minutes, taking short breaks throughout. Unsurprisingly, the participants improved in their performance across the training session; practice, after all, is supposed to make perfect. We then tested the participants twelve hours later. Half of the participants had learned the sequence in the morning and were tested later that evening after remaining awake across the day. The other half of the subjects learned the sequence in the evening and we retested them the next morning after a similar twelve-hour delay, but one that contained a full eight-hour night of sleep.

Those who remained awake across the day showed no evidence of a significant improvement in performance. However, fitting with the pianist's original description, those who were tested after the very same time delay of twelve hours, but that spanned a night of sleep, showed a striking 20 percent jump in performance speed and a near 35 percent improvement in accuracy. Importantly, those participants who learned the motor skill in the morning—and who showed no improvement that evening—did go on to show an identical bump up in performance when retested after a further twelve hours, now after they, too, had had a full night's sleep.

In other words, your brain will continue to improve skill memories in the absence of any further practice. It is really quite magical. Yet, that delayed, "offline" learning occurs exclusively across a period of sleep, and not across equivalent time periods spent awake, regardless of whether the time awake or time asleep comes first. Practice does not make perfect. It is practice, followed by a night of sleep, that leads to perfection. We went on to show that these memory-boosting benefits occur no matter whether you learn a short or a very long motor sequence (e.g., 4-3-1-2 versus 4-2-3-4-2-3-1-4-3-4-1-4), or when using one hand (unimanual) or both (bimanual, similar to a pianist).

Analyzing the individual elements of the motor sequence, such as 4-1-3-2-4, allowed me to discover how, precisely, sleep was perfecting skill. Even after a long period of initial training, participants would consistently struggle with particular transitions within the sequence. These problem points stuck out like a sore thumb when I looked at the speed of the keystrokes. There would be a far longer pause, or consistent error, at specific transitions. For example, rather than seamlessly typing 4-1-3-2-4, 4-1-3-2-4, a participant would instead type: 4-1-3 *[pause]* 2-4, 4-1-3 *[pause]* 2-4. They were chunking the motor routine into pieces, as if attempting the sequences all in one go was just too much. Different people had different pause problems at different points in the routine, but almost all people had one or two of these difficulties. I assessed so many participants that I could actually tell where their unique difficulties in the motor routine were just by listening to their typing during training.

When I tested participants after a night of sleep, however, my ears heard something very different. I knew what was happening even before I analyzed the data: mastery. Their typing, post-sleep, was now fluid and unbroken. Gone was the staccato performance, replaced by seamless automaticity, which is the ultimate goal of motor learning: 4-1-3-2-4, 4-1-3-2-4, 4-1-3-2-4, rapid and nearly perfect. Sleep had systematically identified where the difficult transitions were in the motor memory and smoothed them out. This finding rekindled the words of the pianist I'd met: "but when I wake up the next morning and sit back down at the piano, I can just play, *perfectly*."

I went on to test participants inside a brain scanner after they had slept, and could see how this delightful skill benefit had been achieved. Sleep had again transferred the memories, but the results were different from that for textbook-like memory. Rather than a transfer from short- to long-term memory required for saving facts, the motor memories had been shifted over to brain circuits that operate below the level of consciousness. As a result, those skill actions were now instinctual habits. They flowed out of the body with ease, rather than feeling effortful and deliberate. Which is to say that sleep helped the brain automate the movement routines, making them second nature—effortless—precisely the goal of many an Olympic coach when perfecting the skills of their elite athletes.

My final discovery, in what spanned almost a decade of research, identified the type of sleep responsible for the overnight motor-skill enhancement, carrying with it societal and medical lessons. The increases in speed and accuracy, underpinned by efficient automaticity, were directly related to the amount of stage 2 NREM, especially in the last two hours of an eight-hour night of sleep (e.g., from five to seven a.m., should you have fallen asleep at eleven p.m.). Indeed, it was the number of those wonderful sleep spindles in the last two hours of the late morning—the time of night with the richest spindle bursts of brainwave activity—that were linked with the offline memory boost.

More striking was the fact that the increase of these spindles after learning was detected only in regions of the scalp that sit above the motor cortex (just in front of the crown of your head), and not in other areas. The greater the local increase in sleep spindles over the part of the brain we had forced to learn the motor skill exhaustively, the better the performance upon awakening. Many other groups have found a similar "local-sleep"-and-learning effect. When it comes to motor-skill memories, the brainwaves of sleep were acting like a good masseuse—you still get a full body massage, but they will place special focus on areas of the body that need the most help. In the same way, sleep spindles bathe all parts of the brain, but a disproportionate emphasis will be placed on those parts of the brain that have been worked hardest with learning during the day.

Perhaps more relevant to the modern world is the time-of-night effect we discovered. Those last two hours of sleep are precisely the window that many of us feel it is okay to cut short to get a jump start on the day. As a result, we miss out on this feast of late-morning sleep spindles.

It also brings to mind the prototypical Olympic coach who stoically has her athletes practicing late into the day, only to have them wake in the early hours of the morning and return to practice. In doing so, coaches may be innocently but effectively denying an important phase of motor memory development within the brain—one that fine-tunes skilled athletic performance. When you consider that very small performance differences often separate winning a gold medal from a last-place finish in professional athletics, then any competitive advantage you can gain, such as that naturally offered by sleep, can help determine whether or not you will hear your national anthem echo around the stadium. Not without putting too fine a point on it, if you don't snooze, you lose.

The 100-meter sprint superstar Usain Bolt has, on many occasions, taken naps in the hours before breaking the world record, and before Olympic finals in which he won gold. Our own studies support his wisdom: daytime naps that contain sufficient numbers of sleep spindles also offer significant motor skill memory improvement, together with a restoring benefit on perceived energy and reduced muscle fatigue.

In the years since our discovery, numerous studies have shown that sleep improves the motor skills of junior, amateur, and elite athletes across sports as diverse as tennis, basketball, football, soccer, and rowing. So much so that, in 2015, the International Olympic Committee published a consensus statement highlighting the critical importance of, and essential need for, sleep in athletic development across all sports for men and women.*

Professional sports teams are taking note, and for good reason. I have recently given presentations to a number of national basketball and football teams in the United States, and for the latter, the United Kingdom. Standing in front of the manager, staff, and players, I tell them about one of the most sophisticated, potent, and powerful—not to mention legal—performance enhancers that has real game-winning potential: sleep.

I back up these claims with examples from the more than 750 scientific studies that have investigated the relationship between sleep and

*M. F. Bergeron, M. Mountjoy, N. Armstrong, M. Chia, et al., "International Olympic Committee consensus statement on youth athletic development," *British Journal of Sports Medicine* 49, no. 13 (2015): 843–51.

human performance, many of which have studied professional and elite athletes specifically. Obtain anything less than eight hours of sleep a night, and especially less than six hours a night, and the following happens: time to physical exhaustion drops by 10 to 30 percent, and aerobic output is significantly reduced. Similar impairments are observed in limb extension force and vertical jump height, together with decreases in peak and sustained muscle strength. Add to this marked impairments in cardiovascular, metabolic, and respiratory capabilities that hamper an underslept body, including faster rates of lactic acid buildup, reductions in blood oxygen saturation, and converse increases in blood carbon dioxide, due in part to a reduction in the amount of air that the lungs can expire. Even the ability of the body to cool itself during physical exertion through sweating—a critical part of peak performance—is impaired by sleep loss.

And then there is injury risk. It is the greatest fear of all competitive athletes and their coaches. Concern also comes from the general managers of professional teams, who consider their players as prized financial investments. In the context of injury, there is no better risk-mitigating insurance policy for these investments than sleep. Described in a research study of competitive young athletes in 2014,* you can see that a chronic lack of sleep across the season predicted a markedly higher risk of injury (figure 10). Recent larger-scale analyses have confirmed the same relationship.

Figure 10: Sleep Loss and Sports Injury

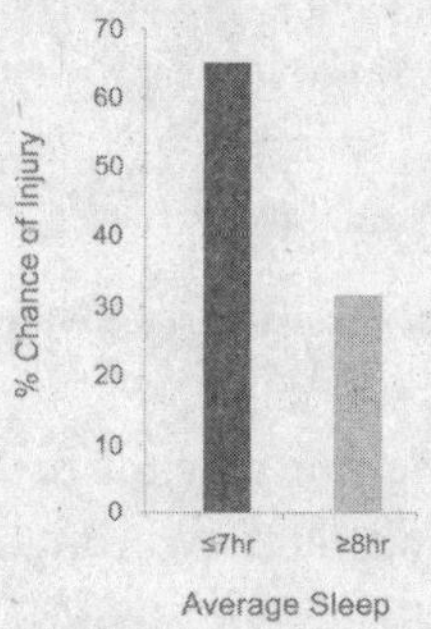

*M. D. Milewski et al., "Chronic lack of sleep is associated with increased sports injuries in adolescent athletes," *Journal of Paediatric Orthopaedics* 34, no. 2 (2014): 129–33.

Sports teams pay millions of dollars to hugely expensive players, lavishing all manner of medical and nutritional care on their human commodities to augment their talent. Yet the professional advantage is diluted several-fold by the one ingredient few teams fail to prioritize: their players' sleep.

Even teams that are aware of sleep's importance before a game are surprised by my declaration of the equally, if not more, essential need for sleep in the days *after* a game. Post-performance sleep accelerates physical recovery from common inflammation, stimulates muscle repair, and helps restock cellular energy in the form of glucose and glycogen.

Prior to giving these teams a structured set of sleep recommendations that they can put in practice to help capitalize on the full potential of their athletes, I provide proof-of-concept data from the National Basketball Association (NBA), using the measured sleep of Andre Iguodala, currently of my local team, the Golden State Warriors. Based on sleep-tracker data, figure 11 is the difference in Iguodala's performance when he's been sleeping more than eight hours a night, relative to less than eight hours a night:*

Figure 11: NBA Player Performance

More than Eight Hours Sleep vs. Less than Eight Hours Sleep

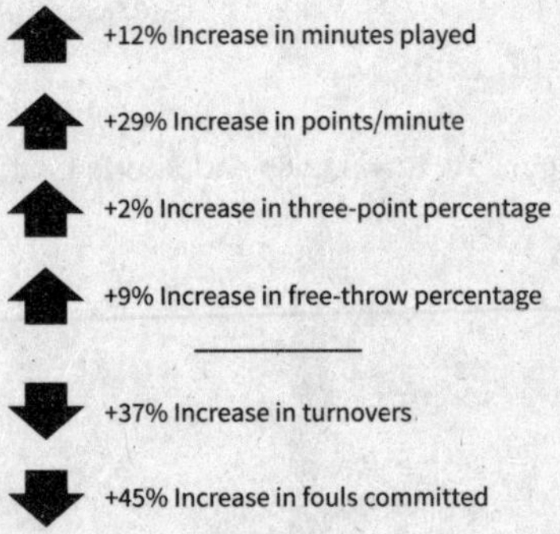

Of course, most of us do not play for professional sports teams. But many of us are physically active throughout life and constantly

*Ken Berger, "In multibillion-dollar business of NBA, sleep is the biggest debt" (CBSSports .com, June 7, 2016).

acquiring new skills. Motor learning and general physicality remain part of our lives, from the banal (learning to type on a slightly new laptop or text on a different-size smartphone) to the essential, such as experienced surgeons learning a new endoscopic procedure or pilots learning to fly different or new aircraft. And so, therefore, we continue to need and rely upon our NREM sleep for refining and maintaining those motor movements. Of interest to parents, the most dramatic time of skilled motor learning in any human's life occurs in the first years after birth, as we start to stand and walk. It is of little surprise that we see a spike in stage 2 NREM sleep, including sleep spindles, right around the infant's time of transition from crawling to walking.

Returning full circle to that which I had learned years ago at the Queen's Medical Center regarding brain damage, we have now discovered that the slow, day-by-day return of motor function in stroke patients is due, in part, to the hard night-by-night work of sleep. Following a stroke, the brain begins to reconfigure those neural connections that remain, and sprout new connections around the damaged zone. This plastic reorganization and the genesis of new connections underlie the return of some degree of motor function. We now have preliminary evidence that sleep is one critical ingredient assisting in this neural recovery effort. Ongoing sleep quality predicts the gradual return of motor function, and further determines the relearning of numerous movement skills.* Should more such findings emerge, then a more concerted effort to prioritize sleep as a therapeutic aid in patients who have suffered brain damage may be warranted, or even the implementation of sleep-stimulation methods like those described earlier. There is much that sleep can do that we in medicine currently cannot. So long as the scientific evidence justifies it, we should make use of the powerful health tool that sleep represents in making our patients well.

---

*K. Herron, D. Dijk, J. Ellis, J. Sanders, and A. M. Sterr, "Sleep correlates of motor recovery in chronic stroke: a pilot study using sleep diaries and actigraphy," *Journal of Sleep Research* 17 (2008): 103; and C. Siengsukon and L. A. Boyd, "Sleep enhances off-line spatial and temporal motor learning after stroke," *Neurorehabilitation & Neural Repair* 4, no. 23 (2009): 327–35.

SLEEP FOR CREATIVITY

A final benefit of sleep for memory is arguably the most remarkable of all: creativity. Sleep provides a nighttime theater in which your brain tests out and builds connections between vast stores of information. This task is accomplished using a bizarre algorithm that is biased toward seeking out the most distant, nonobvious associations, rather like a backward Google search. In ways your waking brain would never attempt, the sleeping brain fuses together disparate sets of knowledge that foster impressive problem-solving abilities. If you ponder the type of conscious experience such outlandish memory blending would produce, you may not be surprised to learn that it happens during the dreaming state—REM sleep. We will fully explore all of the advantages of REM sleep in the later chapter on dreaming. For now, I will simply tell you that such informational alchemy conjured by REM-sleep dreaming has led to some of the greatest feats of transformative thinking in the history of the human race.

CHAPTER 7

Too Extreme for the *Guinness Book of World Records*

Sleep Deprivation and the Brain

Struck by the weight of damning scientific evidence, the *Guinness Book of World Records* has stopped recognizing attempts to break the sleep deprivation world record. Recall that *Guinness* deems it acceptable for a man (Felix Baumgartner) to ascend 128,000 feet into the outer reaches of our atmosphere in a hot-air balloon wearing a spacesuit, open the door of his capsule, stand atop a ladder suspended above the planet, and then free-fall back down to Earth at a top speed of 843 mph (1,358 kmh), passing through the sound barrier while creating a sonic boom with just his body. But the risks associated with sleep deprivation are considered to be far, far higher. Unacceptably high, in fact, based on the evidence.

What is that compelling evidence? In the following two chapters, we will learn precisely why and how sleep loss inflicts such devastating effects on the brain, linking it to numerous neurological and psychiatric conditions (e.g., Alzheimer's disease, anxiety, depression, bipolar disorder, suicide, stroke, and chronic pain), and on every physiological system of the body, further contributing to countless disorders and disease (e.g., cancer, diabetes, heart attacks, infertility, weight gain, obesity, and immune deficiency). No facet of the human body is spared the crippling, noxious harm of sleep loss. We are, as you will see, socially, organizationally, economically, physically, behaviorally, nutritionally, linguistically, cognitively, and emotionally dependent upon sleep.

This chapter deals with the dire and sometimes deadly consequences of inadequate sleep on the brain. The chapter that follows will recount the diverse—though equally ruinous and similarly fatal—effects of short sleep on the body.

PAY ATTENTION

There are many ways in which a lack of sufficient sleep will kill you. Some take time; others are far more immediate. One brain function that buckles under even the smallest dose of sleep deprivation is concentration. The deadly societal consequences of these concentration failures play out most obviously and fatally in the form of drowsy driving. Every hour, someone dies in a traffic accident in the US due to a fatigue-related error.

There are two main culprits of drowsy-driving accidents. The first is people completely falling asleep at the wheel. This happens infrequently, however, and usually requires an individual to be acutely sleep-deprived (having gone without shut-eye for twenty-plus hours). The second, more common cause is a momentary lapse in concentration, called a microsleep. These last for just a few seconds, during which time the eyelid will either partially or fully close. They are usually suffered by individuals who are chronically sleep restricted, defined as getting less than seven hours of sleep a night on a routine basis.

During a microsleep, your brain becomes blind to the outside world for a brief moment—and not just the visual domain, but in all channels of perception. Most of the time you have no awareness of the event. More problematic is that your decisive control of motor actions, such as those necessary for operating a steering wheel or a brake pedal, will momentarily cease. As a result, you don't need to fall asleep for ten to fifteen seconds to die while driving. Two seconds will do it. A two-second microsleep at 30 mph with a modest angle of drift can result in your vehicle transitioning entirely from one lane to the next. This includes into oncoming traffic. Should this happen at 60 mph, it may be the last microsleep you ever have.

David Dinges at the University of Pennsylvania, a titan in the field of sleep research and personal hero of mine, has done more than any scientist in history to answer the following fundamental question: What is the recycle rate of a human being? That is, how long can a human go without sleep before their performance is objectively impaired? How much sleep can a human lose each night, and over how many nights, before critical processes of the brain fail? Is that individual even aware of how impaired they are when sleep-deprived? How many nights of recovery sleep does it take to restore the stable performance of a human after sleep loss?

Dinges's research employs a disarmingly simple attention test to measure concentration. You must press a button in response to a light that appears on a button box or computer screen within a set period of time. Your response, and the reaction time of that response, are both measured. Thereafter, another light comes on, and you do the same thing. The lights appear in an unpredictable manner, sometimes in quick succession, other times randomly separated by a pause lasting several seconds.

Sounds easy, right? Try doing it for ten minutes straight, every day, for fourteen days. That's what Dinges and his research team did to a large number of subjects who were monitored under strict laboratory conditions. All of the subjects started off by getting a full eight-hour sleep opportunity the night before the test, allowing them to be assessed when fully rested. Then, the participants were divided into four different experimental groups. Rather like a drug study, each group was given a different "dose" of sleep deprivation. One group was kept up for seventy-two hours straight, going without sleep for three consecutive nights. The second group was allowed four hours of sleep each night. The third group was given six hours of sleep each night. The lucky fourth group was allowed to keep sleeping eight hours each night.

There were three key findings. First, although sleep deprivation of all these varied amounts caused a slowing in reaction time, there was something more telling: participants would, for brief moments, stop responding altogether. Slowness was not the most sensitive signature of sleepiness, entirely missed responses were. Dinges was capturing lapses, otherwise known as microsleeps: the real-life equivalent of which would be failing to react to a child who runs out in front of your car when chasing a ball.

When describing the findings, Dinges will often have you think of the repeating beep from a heart monitor in a hospital: beep, beep, beep. Now picture the dramatic sound effect you hear in emergency room television dramas when a patient starts to slip away as doctors frantically try to save their life. At first, the heartbeats are constant—beep, beep, beep—as are your responses on the visual attention task when you are well rested: stable, regular. Switch to your performance when sleep-deprived, and it is the aural equivalent of the patient in the hospital going into cardiac arrest: beep, beep, beep, beeeeeeeeeeeeep. Your performance has flatlined. No conscious response, no motor response. A microsleep. And then the heartbeat comes back, as will your performance—beep, beep, beep—but only for a short while. Soon, you have another arrest: beep, beep, beeeeeeeeeeeeep. More microsleeps.

Comparing the number of lapses, or microsleeps, day after day across the four different experimental groups gave Dinges a second key finding. Those individuals who slept eight hours every night maintained a stable, near-perfect performance across the two weeks. Those in the three-night total sleep deprivation group suffered catastrophic impairment, which was no real surprise. After the first night of no sleep at all, their lapses in concentration (missed responses) increased by over 400 percent. The surprise was that these impairments continued to escalate at the same ballistic rate after a second and third night of total sleep deprivation, as if they would continue to escalate in severity if more nights of sleep were lost, showing no signs of flattening out.

But it was the two partial sleep deprivation groups that brought the most concerning message of all. After four hours of sleep for six nights, participants' performance was just as bad as those who had not slept for twenty-four hours straight—that is, a 400 percent increase in the number of microsleeps. By day 11 on this diet of four hours of sleep a night, participants' performance had degraded even further, matching that of someone who had pulled two back-to-back all-nighters, going without sleep for forty-eight hours.

Most worrying from a societal perspective were the individuals in the group who obtained six hours of sleep a night—something that may sound familiar to many of you. Ten days of six hours of sleep a night was all it took to become as impaired in performance as going without sleep for twenty-four hours straight. And like the total sleep deprivation group, the accruing performance impairment in the four-hour and six-hour sleep groups showed no signs of leveling out. All signs suggested that if the experiment had continued, the performance deterioration would continue to build up over weeks or months.

Another research study, this one led by Dr. Gregory Belenky at Walter Reed Army Institute of Research, published almost identical results around the same time. They also tested four groups of participants, but they were given nine hours, seven hours, five hours, and three hours of sleep across seven days.

YOU DO NOT KNOW HOW SLEEP-DEPRIVED YOU ARE WHEN YOU ARE SLEEP-DEPRIVED

The third key finding, common to both of these studies, is the one I personally think is the most harmful of all. When participants were asked about their subjective sense of how impaired they were, they consistently underestimated their degree of performance disability. It was a miserable predictor of how bad their performance actually, objectively was. It is the equivalent of someone at a bar who has had far too many drinks picking up his car keys and confidently telling you, "I'm fine to drive home."

Similarly problematic is baseline resetting. With chronic sleep restriction over months or years, an individual will actually acclimate to their impaired performance, lower alertness, and reduced energy levels. That low-level exhaustion becomes their accepted norm, or baseline. Individuals fail to recognize how their perennial state of sleep deficiency has come to compromise their mental aptitude and physical vitality, including the slow accumulation of ill health. A link between the former and latter is rarely made in their mind. Based on epidemiological studies of average sleep time, millions of individuals unwittingly spend years of their life in a sub-optimal state of psychological and physiological functioning, never maximizing their potential of mind or body due to their blind persistence in sleeping too little. Sixty years of scientific research prevent me from accepting anyone who tells me that he or she can "get by on just four or five hours of sleep a night just fine."

Returning to Dinges's study results, you may have predicted that optimal performance would return to all of the participants after a good long night of recovery sleep, similar to many people's notion of "sleeping it off" on the weekends to pay off their weeknight sleep debt. However, even after three nights of ad lib recovery sleep, performance did not return to that observed at the original baseline assessment when those same individuals had been getting a full eight hours of sleep regularly. Nor did any group recover all the sleep hours they had lost in the days prior. As we have already learned, the brain is incapable of that.

In a disturbing later study, researchers in Australia took two groups of healthy adults, one of whom they got drunk to the legal driving limit (.08 percent blood alcohol), the other of whom they sleep-deprived for a single night. Both groups performed the concentration test to assess attention performance, specifically the number of lapses. After being awake for nineteen hours, people who were sleep-deprived were as cognitively impaired as those who were legally drunk. Said another way, if you wake up at seven a.m. and remain awake throughout the day, then go out socializing with friends until late that evening, yet drink no alcohol whatsoever, by the time you are driving home at two a.m. you are as cognitively impaired in your ability to attend to the road and what is around you as a legally drunk driver. In fact, participants in the above study started their nosedive in performance after just fifteen hours of being awake (ten p.m. in the above scenario).

Car crashes rank among the leading causes of death in most first-world nations. In 2016, the AAA Foundation in Washington, DC, released the results of an extensive study of over 7,000 drivers in the US, tracked in detail over a two-year period.* The key finding, shown in figure 12, reveals just how catastrophic drowsy driving is when it comes to car crashes. Operating on less than five hours of sleep, your risk of a car crash increases threefold. Get behind the wheel of a car when having slept just four hours or less the night before and you are 11.5 times more likely to be involved in a car accident. Note how the relationship between decreasing hours of sleep and increasing mortality risk of an

*Foundation for Traffic Safety. "Acute Sleep Deprivation and Crash Risk," accessed at https://www.aaafoundation.org/acute-sleep-deprivation-and-crash-risk.

accident is not linear, but instead exponentially mushrooms. Each hour of sleep lost vastly amplifies that crash likelihood, rather than incrementally nudging it up.

Figure 12: Sleep Loss and Car Crashes

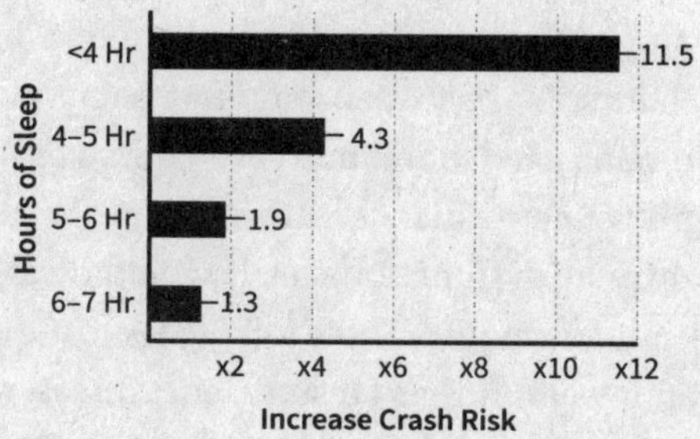

Drunk driving and drowsy driving are deadly propositions in their own right, but what happens when someone combines them? It is a relevant question, since most individuals are driving drunk in the early-morning hours rather than in the middle of the day, meaning that most drunk drivers are also sleep-deprived.

We can now monitor driver error in a realistic but safe way using driving simulators. With such a virtual machine, a group of researchers examined the number of complete off-road deviations in participants placed under four different experimental conditions: (1) eight hours of sleep, (2) four hours of sleep, (3) eight hours of sleep plus alcohol to the point of being legally drunk, and (4) four hours of sleep plus alcohol to the point of being legally drunk.

Those in the eight-hour sleep group had few, if any, off-road errors. Those in the four-hour sleep condition (the second group) had six times more off-road deviations than the sober, well-rested individuals. The same degree of driving impairment was true of the third group, who had eight hours of sleep but were legally drunk. Driving drunk or driving drowsy were both dangerous, and equally dangerous.

A reasonable expectation was that performance in the fourth group of participants would reflect the additive impact of these two groups: four hours of sleep plus the effect of alcohol (i.e., twelve times more off-road deviations). It was far worse. This group of participants drove off the road almost thirty times more than the well-rested, sober group.

The heady cocktail of sleep loss and alcohol was not *additive*, but instead *multiplicative*. They magnified each other, like two drugs whose effects are harmful by themselves but, when taken together, interact to produce truly dire consequences.

After thirty years of intensive research, we can now answer many of the questions posed earlier. The recycle rate of a human being is around sixteen hours. After sixteen hours of being awake, the brain begins to fail. Humans need more than seven hours of sleep each night to maintain cognitive performance. After ten days of just seven hours of sleep, the brain is as dysfunctional as it would be after going without sleep for twenty-four hours. Three full nights of recovery sleep (i.e., more nights than a weekend) are insufficient to restore performance back to normal levels after a week of short sleeping. Finally, the human mind cannot accurately sense how sleep-deprived it is when sleep-deprived.

We shall return to the ramifications of these results in the remaining chapters, but the real-life consequences of drowsy driving deserve special mention. This coming week, more than 2 million people in the US will fall asleep while driving their motor vehicle. That's more than 250,000 every day, with more such events during the week than weekends for obvious reasons. More than 56 million Americans admit to struggling to stay awake at the wheel of a car each month.

Drowsy driving alone may be as, if not more, problematic than driving drunk: drunk drivers are often *late* in braking, and *late* in making evasive maneuvers. But when you fall asleep, or have a microsleep, *you stop reacting altogether*. A person who experiences a microsleep or who has fallen asleep at the wheel does not brake at all, nor do they make any attempt to avoid the accident. As a result, car crashes caused by drowsiness tend to be far more deadly than those caused by alcohol or drugs. Said crassly, when you fall asleep at the wheel of your car on a freeway, there is now a one-ton missile traveling at 65 miles per hour, and no one is in control.

Drivers of cars are not the only threats. More dangerous are drowsy truckers. Approximately 80 percent of truck drivers in the US are overweight, and 50 percent are clinically obese. This places truck drivers at a far, far higher risk of a disorder called sleep apnea, commonly associated with heavy snoring, which causes chronic, severe sleep deprivation. As a result, these truck drivers are far more likely to be involved in a traffic accident. And when a truck driver loses his or her life in a drowsy-driving crash, they can take a number of lives with them.

In actual fact, I would like to argue that there are no *accidents* caused by fatigue, microsleeps, or falling asleep. None whatsoever. They are *crashes*. The *Oxford English Dictionary* defines accidents as unexpected events that happen by chance or without apparent cause. Drowsy-driving deaths are neither chance, nor without cause. They are predictable and the direct result of not obtaining sufficient sleep. As such, they are unnecessary and preventable. Shamefully, governments of most developed countries spend less than 1 percent of their budget educating the public on the dangers of drowsy driving relative to what they invest in combating drunk driving.

Even well-meaning public health messages can get lost in a barrage of statistics. It often takes the tragic recounting of personal stories to make the message real. There are thousands of such events that I could describe. Let me offer just one in the hopes of saving you from the harms of driving drowsy.

Union County, Florida, January 2006: a school bus transporting nine children came to a halt at a stop sign. A Pontiac Bonneville car carrying seven occupants pulled up behind the bus and also came to a stop. At this moment, an eighteen-wheel truck came barreling down the road behind both vehicles. It didn't stop. The truck struck the Pontiac, riding up over it and, with the car concertinaed underneath, then hit the bus. All three vehicles traveled through a ditch and continued moving, at which point the imploded Pontiac became engulfed in flames. The school bus rotated counterclockwise and kept traveling, now on the opposite side of the road, back-to-front. It did so for 328 feet until it went off the road and collided with a thick grove of trees. Three of the nine children in the bus were ejected through the windows upon impact. All seven passengers in the Pontiac were killed, as was the bus driver. The truck driver and all nine children in the bus sustained serious injuries.

The trucker was a qualified and legally licensed driver. All toxicology tests performed on his blood were negative. However, it later emerged that he had been awake for thirty-four hours straight and had fallen asleep at the wheel. All of the Pontiac's seven occupants who died were children or adolescents. Five of the seven children in the Pontiac car were from a single family. The oldest occupant was a teenager, who had been legally driving the car. The youngest occupant was a baby of just twenty months old.

There are many things that I hope readers take away from this book. This is one of the most important: if you are drowsy while driving, *please, please stop*. It is lethal. To carry the burden of another's death on your shoulders is a terrible thing. Don't be misled by the many ineffective tactics people will tell you can battle back against drowsiness while driving.* Many of us think we can overcome drowsiness through sheer force of will, but, sadly, this is not true. To assume otherwise can jeopardize your life, the lives of your family or friends in the car with you, and the lives of other road users. Some people only get one chance to fall asleep at the wheel before losing their life.

If you notice yourself feeling drowsy while driving, or actually falling asleep at the wheel, stop for the night. If you really must keep going—and you have made that judgment in the life-threatening context it genuinely poses—then pull off the road into a safe layby for a short time. Take a brief nap (twenty to thirty minutes). When you wake up, *do not* start driving. You will be suffering from sleep inertia—the carry-over effects of sleep into wakefulness. Wait for another twenty to thirty minutes, perhaps after having a cup of coffee if you really must, and only then start driving again. This, however, will only get you so far down the road before you need another such recharge, and the returns are diminishing. Ultimately, it is just not worth the (life) cost.

CAN NAPS HELP?

In the 1980s and '90s, David Dinges, together with his astute collaborator (and recent administrator of the National Highway Traffic Safety

*Common myths that are of no use in helping to overcome drowsiness while driving include: turning up the radio, winding down the car window, blowing cold air on your face, splashing cold water on your face, talking on the phone, chewing gum, slapping yourself, pinching yourself, punching yourself, and promising yourself a reward for staying awake.

Administration) Dr. Mark Rosekind, conducted another series of groundbreaking studies, this time examining the upsides and downsides of napping in the face of unavoidable sleep deprivation. They coined the term "power naps"—or, should I say, ceded to it. Much of their work was with the aviation industry, examining pilots on long-haul travel.

The most dangerous time of flight is landing, which arrives at the end of a journey, when the greatest amount of sleep deprivation has often accrued. Recall how tired and sleepy you are at the end of an overnight, transatlantic flight, having been on the go for more than twenty-four hours. Would you feel at peak performance, ready to land a Boeing 747 with 467 passengers on board, should you have the skill to do so? It is during this end phase of flight, known in the aviation industry as "top of descent to landing," that 68 percent of all hull losses—a euphemism for a catastrophic plane crash—occur.

The researchers set to work answering the following question, posed by the US Federal Aviation Authority (FAA): If a pilot can only obtain a short nap opportunity (40–120 minutes) within a thirty-six-hour period, when should it occur so as to minimize cognitive fatigue and attention lapses: at the start of the first evening, in the middle of the night, or late the following morning?

It first appeared to be counterintuitive, but Dinges and Rosekind made a clever, biology-based prediction. They believed that by inserting a nap at the front end of an incoming bout of sleep deprivation, you could insert a buffer, albeit temporary and partial, that would protect the brain from suffering catastrophic lapses in concentration. They were right. Pilots suffered fewer microsleeps at the end stages of the flight if the naps were taken early that prior evening, versus if those same nap periods were taken in the middle of the night or later that next morning, when the attack of sleep deprivation was already well under way.

They had discovered the sleep equivalent of the medical paradigm of prevention versus treatment. The former tries to avert an issue prior to occurrence, the latter tries to remedy the issue after it has happened. And so it was with naps. Indeed, these short sleep bouts, taken early, also reduced the number of times the pilots drifted into light sleep during the critical, final ninety minutes of flight. There were fewer of these sleep intrusions, measured with EEG electrodes on the head.

When Dinges and Rosekind reported their findings to the FAA, they recommended that "prophylactic naps"—naps taken early during long-haul flights—should be instituted as policy among pilots, as many other aviation authorities around the world now permit. The FAA, while believing the findings, was not convinced by the nomenclature. They believed the term "prophylactic" was ripe for many a snide joke among pilots. Dinges suggested the alternative of "planned napping." The FAA didn't like this, either, feeling it to be too "management-like." Their suggestion was "power napping," which they believed was more fitting with leadership- or dominance-based job positions, others being CEOs or military executives. And so the "power nap" was born.

The problem, however, is that people, especially those in such positions, came to erroneously believe that a twenty-minute power nap was all you needed to survive and function with perfect, or even acceptable, acumen. Brief power naps have become synonymous with the inaccurate assumption that they allow an individual to forgo sufficient sleep, night after night, especially when combined with the liberal use of caffeine.

No matter what you may have heard or read in the popular media, there is no scientific evidence we have suggesting that a drug, a device, or any amount of psychological willpower can replace sleep. Power naps may momentarily increase basic concentration under conditions of sleep deprivation, as can caffeine up to a certain dose. But in the subsequent studies that Dinges and many other researchers (myself included) have performed, neither naps nor caffeine can salvage more complex functions of the brain, including learning, memory, emotional stability, complex reasoning, or decision-making.

One day we may discover such a counteractive method. Currently, however, there is no drug that has the proven ability to replace those benefits that a full night of sleep infuses into the brain and body. David Dinges has extended an open invitation to anyone suggesting that they can survive on short sleep to come to his lab for a ten-day stay. He will place that individual on their proclaimed regiment of short sleep and measure their cognitive function. Dinges is rightly confident he'll show, categorically, a degradation of brain and body function. To date, no volunteers have matched up to their claim.

We have, however, discovered a very rare collection of individuals who appear to be able to survive on six hours of sleep, and show minimal impairment—a sleepless elite, as it were. Give them hours and hours of sleep opportunity in the laboratory, with no alarms or wake-up calls, and still they naturally sleep this short amount and no more. Part of the explanation appears to lie in their genetics, specifically a sub-variant of a gene called BHLHE41.* Scientists are now trying to understand what this gene does, and how it confers resilience to such little sleep.

Having learned this, I imagine that some readers now believe that they are one of these individuals. That is very, very unlikely. The gene is remarkably rare, with but a soupçon of individuals in the world estimated to carry this anomaly. To impress this fact further, I quote one of my research colleagues, Dr. Thomas Roth at the Henry Ford Hospital in Detroit, who once said, "The number of people who can survive on five hours of sleep or less without any impairment, expressed as a percent of the population, and rounded to a whole number, is zero."

There is but a fraction of 1 percent of the population who are truly resilient to the effects of chronic sleep restriction at all levels of brain function. It is far, far more likely that you will be struck by lightning (the lifetime odds being 1 in 12,000) than being truly capable of surviving on insufficient sleep thanks to a rare gene.

EMOTIONAL IRRATIONALITY

"I just snapped, and . . ." Those words are often part of an unfolding tragedy as a soldier irrationally responds to a provocative civilian, a physician to an entitled patient, or a parent to a misbehaving child. All of these situations are ones in which inappropriate anger and hostility are dealt out by tired, sleep-deprived individuals.

Many of us know that inadequate sleep plays havoc with our emo-

*Also known as DEC2.

tions. We even recognize it in others. Consider another common scenario of a parent holding a young child who is screaming or crying and, in the midst of the turmoil, turns to you and says, "Well, Steven just didn't get enough sleep last night." Universal parental wisdom knows that bad sleep the night before leads to a bad mood and emotional reactivity the next day.

While the phenomenon of emotional irrationality following sleep loss is subjectively and anecdotally common, until recently we did not know how sleep deprivation influenced the emotional brain at a neural level, despite the professional, psychiatric, and societal ramifications. Several years ago, my team and I conducted a study using MRI brain scanning to address the question.

We studied two groups of healthy young adults. One group stayed awake all night, monitored under full supervision in my laboratory, while the other group slept normally that night. During the brain scanning session the next day, participants in both groups were shown the same one hundred pictures that ranged from neutral in emotional content (e.g., a basket, a piece of driftwood) to emotionally negative (e.g., a burning house, a venomous snake about to strike). Using this emotional gradient of pictures, we were able to compare the increase in brain response to the increasingly negative emotional triggers.

Analysis of the brain scans revealed the largest effects I have measured in my research to date. A structure located in the left and right sides of the brain, called the amygdala—a key hot spot for triggering strong emotions such as anger and rage, and linked to the fight-or-flight response—showed well over a 60 percent amplification in emotional reactivity in the participants who were sleep-deprived. In contrast, the brain scans of those individuals who were given a full night's sleep evinced a controlled, modest degree of reactivity in the amygdala, despite viewing the very same images. It was as though, without sleep, our brain reverts to a primitive pattern of uncontrolled reactivity. We produce unmetered, inappropriate emotional reactions, and are unable to place events into a broader or considered context.

This answer raised another question: Why were the emotion centers of the brain so excessively reactive without sleep? Further MRI studies using more refined analyses allowed us to identify the root cause.

After a full night of sleep, the prefrontal cortex—the region of the brain that sits just above your eyeballs; is most developed in humans, relative to other primates; and is associated with rational, logical thought and decision-making—was strongly coupled to the amygdala, regulating this deep emotional brain center with inhibitory control. With a full night of plentiful sleep, we have a balanced mix between our emotional gas pedal (amygdala) and brake (prefrontal cortex). Without sleep, however, the strong coupling between these two brain regions is lost. We cannot rein in our atavistic impulses—too much emotional gas pedal (amygdala) and not enough regulatory brake (prefrontal cortex). Without the rational control given to us each night by sleep, we're not on a neurological—and hence emotional—even keel.

Recent studies by a research team in Japan have now replicated our findings, but they've done so by restricting participants' sleep to five hours for five nights. No matter how you take sleep from the brain—acutely, across an entire night, or chronically, by short sleeping for a handful of nights—the emotional brain consequences are the same.

When we conducted our original experiments, I was struck by the pendulum-like swings in the mood and emotions of our participants. In a flash, sleep-deprived subjects would go from being irritable and antsy to punch-drunk giddy, only to then swing right back to a state of vicious negativity. They were traversing enormous emotional distances, from negative to neutral to positive, and all the way back again, within a remarkably short period of time. It was clear that I had missed something. I needed to conduct a sister study to the one I described above, but now explore how the sleep-deprived brain responds to increasingly positive and rewarding experiences, such as exciting images of extreme sports, or the chance of winning increasing amounts of money in fulfilling tasks.

We discovered that different deep emotional centers in the brain just above and behind the amygdala, called the striatum—associated with impulsivity and reward, and bathed by the chemical dopamine—had become hyperactive in sleep-deprived individuals in response to the rewarding, pleasurable experiences. As with the amygdala, the heightened sensitivity of these hedonic regions was linked to a loss of the rational control from the prefrontal cortex.

Insufficient sleep does not, therefore, push the brain into a negative mood state and hold it there. Rather, the under-slept brain swings excessively to both extremes of emotional valence, positive and negative.

You may think that the former counter-balances the latter, thereby neutralizing the problem. Sadly, emotions, and their guiding of optimal decision and actions, do not work this way. Extremity is often dangerous. Depression and extreme negative mood can, for example, infuse an individual with a sense of worthlessness, together with ideas of questioning life's value. There is now clearer evidence of this concern. Studies of adolescents have identified a link between sleep disruption and suicidal thoughts, suicide attempts, and, tragically, suicide completion in the days after. One more reason for society and parents to value plentiful sleep in teens rather than chastise it, especially considering that suicide is the second-leading cause of death in young adults in developed nations after car accidents.

Insufficient sleep has also been linked to aggression, bullying, and behavioral problems in children across a range of ages. A similar relationship between a lack of sleep and violence has been observed in adult prison populations; places that, I should add, are woefully poor at enabling good sleep that could reduce aggression, violence, psychiatric disturbance, and suicide, which, beyond the humanitarian concern, increases costs to the taxpayer.

Equally problematic issues arise from extreme swings in positive mood, though the consequences are different. Hypersensitivity to pleasurable experiences can lead to sensation-seeking, risk-taking, and addiction. Sleep disturbance is a recognized hallmark associated with addictive substance use.* Insufficient sleep also determines relapse rates in numerous addiction disorders, associated with reward

*K. J. Brower and B. E. Perron, "Sleep disturbance as a universal risk factor for relapse in addictions to psychoactive substances," *Medical Hypotheses* 74, no. 5 (2010): 928–33; D. A. Ciraulo, J. Piechniczek-Buczek, and E. N. Iscan, "Outcome predictors in substance use disorders," *Psychiatric Clinics of North America* 26, no. 2 (2003): 381–409; J. E. Dimsdale, D. Norman, D. DeJardin, and M. S. Wallace, "The effect of opioids on sleep architecture," *Journal of Clinical Sleep Medicine* 3, no. 1 (2007): 33–36; E. F. Pace-Schott, R. Stickgold, A. Muzur, P. E. Wigren, et al., "Sleep quality deteriorates over a binge-abstinence cycle in chronic smoked cocaine users," *Psychopharmacology (Berl)* 179, no. 4 (2005): 873–83; and J. T. Arnedt, D. A. Conroy, and K. J. Brower, "Treatment options for sleep disturbances during alcohol recovery," *Journal of Addictive Diseases* 26, no. 4 (2007): 41–54.

cravings that are unmetered, lacking control from the rational head office of the brain's prefrontal cortex.* Relevant from a prevention standpoint, insufficient sleep during childhood significantly predicts early onset of drug and alcohol use in that same child during their later adolescent years, even when controlling for other high-risk traits, such as anxiety, attention deficits, and parental history of drug use.† You can now appreciate why the bidirectional, pendulum-like emotional liability caused by sleep deprivation is so concerning, rather than counter-balancing.

Our brain scanning experiments in healthy individuals offered reflections on the relationship between sleep and psychiatric illnesses. There is no major psychiatric condition in which sleep is normal. This is true of depression, anxiety, post-traumatic stress disorder (PTSD), schizophrenia, and bipolar disorder (once known as manic depression).

Psychiatry has long been aware of the coincidence between sleep disturbance and mental illness. However, a prevailing view in psychiatry has been that mental disorders cause sleep disruption—a one-way street of influence. Instead, we have demonstrated that otherwise healthy people can experience a neurological pattern of brain activity similar to that observed in many of these psychiatric conditions simply by having their sleep disrupted or blocked. Indeed, many of the brain regions commonly impacted by psychiatric mood disorders are the same regions that are involved in sleep regulation and impacted by sleep loss. Further, many of the genes that show abnormalities in psychiatric illnesses are the same genes that help control sleep and our circadian rhythms.

Had psychiatry got the causal direction wrong, and it was sleep disruption instigating mental illness, not the other way around? No, I believe that is equally inaccurate and reductionist to suggest. Instead, I firmly believe that sleep loss and mental illness is best described as a two-way street of interaction, with the flow of traffic being stronger in one direction or the other, depending on the disorder.

---

*K. J. Brower and B. E. Perron, "Sleep disturbance as a universal risk factor for relapse in addictions to psychoactive substances," *Medical Hypotheses* 74, no. 5 (2010): 928–33.

†N. D. Volkow, D. Tomasi, G. J. Wang, F. Telang, et al., "Hyperstimulation of striatal D2 receptors with sleep deprivation: Implications for cognitive impairment," *NeuroImage* 45, no. 4 (2009): 1232–40.

I am not suggesting that all psychiatric conditions are caused by absent sleep. However, I am suggesting that sleep disruption remains a neglected factor contributing to the instigation and/or maintenance of numerous psychiatric illnesses, and has powerful diagnostic and therapeutic potential that we are yet to fully understand or make use of.

Preliminary (but compelling) evidence is beginning to support this claim. One example involves bipolar disorder, which most people will recognize by the former name of manic depression. Bipolar disorder should not be confused with major depression, in which individuals slide exclusively down into the negative end of the mood spectrum. Instead, patients with bipolar depression vacillate between both ends of the emotion spectrum, experiencing dangerous periods of mania (excessive, reward-driven emotional behavior) and also periods of deep depression (negative moods and emotions). These extremes are often separated by a time when the patients are in a stable emotional state, neither manic nor depressed.

A research team in Italy examined bipolar patients during the time when they were in this stable, inter-episode phase. Next, under careful clinical supervision, they sleep-deprived these individuals for one night. Almost immediately, a large proportion of the individuals either spiraled into a manic episode or became seriously depressed. I find it to be an ethically difficult experiment to appreciate, but the scientists had importantly demonstrated that a lack of sleep is a causal trigger of a psychiatric episode of mania or depression. The result supports a mechanism in which the sleep disruption—which almost always precedes the shift from a stable to an unstable manic or depressive state in bipolar patients—may well be a (the) trigger in the disorder, and not simply epiphenomenal.

Thankfully, the opposite is also true. Should you improve sleep quality in patients suffering from several psychiatric conditions using a technique we will discuss later, called cognitive behavioral therapy for insomnia (CBT-I), you can improve symptom severity and remission rates. My colleague at the University of California, Berkeley, Dr. Allison Harvey has been a pioneer in this regard.

By improving sleep quantity, quality, and regularity, Harvey and her team have systematically demonstrated the healing abilities of sleep for the minds of numerous psychiatric populations. She has intervened with the therapeutic tool of sleep in conditions as diverse as depression, bipolar disorder, anxiety, and suicide, all to great effect. By regularizing and enhancing sleep, Harvey has stepped these patients back from the edge of crippling mental illness. That, in my opinion, is a truly remarkable service to humanity.

The swings in emotional brain activity that we observed in healthy individuals who were sleep-deprived may also explain a finding that has perplexed psychiatry for decades. Patients suffering from major depression, in which they become exclusively locked into the negative end of the mood spectrum, show what at first appears to be a counterintuitive response to one night of sleep deprivation. Approximately 45 percent of these patients will feel *better* after a night without sleep. Their lack of slumber appears to be an antidepressant.

The reason sleep deprivation is not a commonly used treatment, however, is that as soon as these individuals do sleep, the antidepressant benefit typically goes away. As a result, sleep deprivation alone is not a realistic or comprehensive therapy option. Still, it has posed an interesting question: How could sleep deprivation prove helpful for some of these individuals, yet detrimental to others?

I believe that one explanation resides in the bidirectional changes in emotional brain activity that we observed. Depression is not, as you may think, just about the excess presence of negative feelings. Major depression has as much to do with *absence* of positive emotions, a feature described as anhedonia: the inability to gain pleasure from normally pleasurable experiences, such as food, socializing, or sex.

The one-third of depressed individuals who respond to sleep deprivation may therefore be those who experience the greater amplification within reward circuits of the brain that I described earlier, resulting in far stronger sensitivity to, and experiencing of, positive rewarding triggers following sleep deprivation. Their anhedonia is therefore lessened, and now they can begin to experience a greater degree of pleasure from pleasurable life experiences. In contrast, the other two-thirds of depressed patients may suffer the opposite negative emotional consequences of sleep deprivation more dominantly: a worsening, rather than alleviation, of their depression. If we can identify what determines those who will be responders and those who will not, my hope is that we can create better, more tailored sleep-intervention methods for combating depression.

We will revisit the effects of sleep loss on emotional stability and other brain functions in later chapters when we discuss the real-life consequences of sleep loss in society, education, and the workplace. The findings justify our questioning of whether or not sleep-deprived doctors can make emotionally rational decisions and judgments; under-slept military personnel should have their fingers on the triggers of weaponry; overworked bankers and stock traders can make rational, non-risky financial decisions when investing the public's hard-earned retirement funds; and if teenagers should be battling against impossibly early start times during a developmental phase of life when they are most vulnerable to developing psychiatric disorders. For now, however, I will summarize this section by offering a discerning quote on the topic of sleep and emotion by the American entrepreneur E. Joseph Cossman: "The best bridge between despair and hope is a good night's sleep."*

TIRED AND FORGETFUL?

Have you ever pulled an "all-nighter," deliberately staying awake all night? One of my true loves is teaching a large undergraduate class on the science of sleep at the University of California, Berkeley. I taught a similar sleep course while I was at Harvard University. At the start of the course, I conduct a sleep survey, inquiring about my students' sleep habits, such as the times they go to bed and wake up during the week and weekend, how much sleep they get, if they think their academic performance is related to their sleep.

Inasmuch as they are telling me the truth (they fill the survey out anonymously online, not in class), the answer I routinely get is saddening. More than 85 percent of them pull all-nighters. Especially concerning is the fact that of those who said "yes" to pulling all-nighters, almost a third will do so monthly, weekly, or even several times a week. As the course continues throughout the semester, I return to the results

*Cossman had other pearls of wisdom, too, such as "The best way to remember your wife's birthday is to forget it once."

of their sleep survey and link their own sleep habits with the science we are learning about. In this way, I try to point out the very personal dangers they face to their psychological and physical health due to their insufficient sleep, and the danger they themselves pose to society as a consequence.

The most common reason my students give for pulling all-nighters is to cram for an exam. In 2006, I decided to conduct an MRI study to investigate whether they were right or wrong to do so. Was pulling an all-nighter a wise idea for learning? We took a large group of individuals and assigned them to either a sleep group or a sleep deprivation group. Both groups remained awake normally across the first day. Across the following night, those in the sleep group obtained a full night of shut-eye, while those in the sleep deprivation group were kept awake all night under the watchful eye of trained staff in my lab. Both groups were then awake across the following morning. Around midday, we placed participants inside an MRI scanner and had them try to learn a list of facts, one at a time, as we took snapshots of their brain activity. Then we tested them to see how effective that learning had been. However, instead of testing them immediately after learning, we waited until they had had two nights of recovery sleep. We did this to make sure that any impairments we observed in the sleep-deprived group were not confounded by them being too sleepy or inattentive to recollect what they may very well have learned. Therefore, the sleep-deprivation manipulation was only in effect during the act of learning, and not during the later act of recall.

When we compared the effectiveness of learning between the two groups, the result was clear: there was a 40 percent deficit in the ability of the sleep-deprived group to cram new facts into the brain (i.e., to make new memories), relative to the group that obtained a full night of sleep. To put that in context, it would be the difference between acing an exam and failing it miserably!

What was going wrong within the brain to produce these deficits? We compared the patterns of brain activity during attempted learning between the two groups, and focused our analysis on the brain region that we spoke about in chapter 6, the hippocampus—the information "in-box" of the brain that acquires new facts. There was lots of healthy, learning-related activity in the hippocampus in the participants who had slept the night before. However, when we looked at this same brain structure in the sleep-deprived participants, we could not find any significant learning activity whatsoever. It was as though sleep deprivation had shut down their memory in-box, and any new incoming information was simply being bounced. You don't even need the blunt force of a whole night of sleep deprivation. Simply disrupting the depth of an individual's NREM sleep with infrequent sounds, preventing deep sleep and keeping the brain in shallow sleep, without waking the individual up will produce similar brain deficits and learning impairments.

You may have seen a movie called *Memento*, in which the lead character suffers brain damage and, from that point forward, can no longer make any new memories. In neurology, he is what we call "densely amnesic." The part of his brain that was damaged was the hippocampus. It is the very same structure that sleep deprivation will attack, blocking your brain's capacity for new learning.

I cannot tell you how many of my students have come up to me at the end of the lecture in which I describe these studies and said, "I know that exact feeling. It seems as though I'm staring at the page of the textbook but nothing is going in. I may be able to hold on to some facts the following day for the exam, but if you were to ask me to take that same test a month later, I think I'd hardly remember a thing."

The latter description has scientific backing. Those few memories you are able to learn while sleep-deprived are forgotten far more quickly in the hours and days thereafter. Memories formed without sleep are weaker memories, evaporating rapidly. Studies in rats have found that it is almost impossible to strengthen the synaptic connections between individual neurons that normally forge a new memory circuit in the animals that have been sleep-deprived. Imprinting lasting memories into the architecture of the brain becomes nearly impossible. This is true whether the researchers sleep-deprived the rats for a full twenty-four hours, or just a little, for two or three hours. Even the most elemental units of the learning process—the production of proteins that form the building blocks of memories within these synapses—are stunted by the state of sleep loss.

The very latest work in this area has revealed that sleep deprivation even impacts the DNA and the learning-related genes in the brain cells of the hippocampus itself. A lack of sleep therefore is a deeply penetrating and corrosive force that enfeebles the memory-making apparatus within your brain, preventing you from constructing lasting memory traces. It is rather like building a sand castle too close to the tide line—the consequences are inevitable.

While at Harvard University, I was invited to write my first op-ed piece for their newspaper, the *Crimson.* The topic was sleep loss, learning, and memory. It was also the last piece I was invited to write.

In the article, I described the above studies and their relevance, returning time and again to the pandemic of sleep deprivation that was sweeping through the student body. However, rather than lambaste the students for these practices, I pointed a scolding finger directly at the faculty, myself included. I suggested that if we, as teachers, strive to accomplish just that purpose—to teach—then end-loading exams in the final days of the semester was an asinine decision. It forced a behavior in our students—that of short sleeping or pulling all-nighters leading up to the exam—that was in direct opposition to the goals of nurturing young scholarly minds. I argued that logic, backed by scientific fact, must prevail, and that it was long past the time for us to rethink our evaluation methods, their contra-educational impact, and the unhealthy behavior it coerced from our students.

To suggest that the reaction from the faculty was icy would be a thermal compliment. "It was the students' choice," I was told in adamant response emails. "A lack of planned study by irresponsible undergraduates" was another common rebuttal from faculty and administrators attempting to sidestep responsibility. In truth, I never believed that one op-ed column would trigger a U-turn in poor educational examination methods at that or any other higher institute of learning. As many have said about such stoic institutions: theories, beliefs, and practices die one generation at a time. But the conversation and battle must start somewhere.

You may ask whether I have changed my own educational practice and assessment. I have. There are no "final" exams at the end of the semester in my classes. Instead, I split my courses up into thirds so that students only have to study a handful of lectures at a time. Furthermore, none of the exams are cumulative. It's a tried-and-true effect in the psychology of memory, described as mass versus spaced learning. As with a fine-dining experience, it is far more preferable to separate the educational meal into smaller courses, with breaks in between to allow for digestion, rather than attempt to cram all of those informational calories down in one go.

In chapter 6 I described the crucial role for sleep after learning in the offline cementing, or consolidating, of recently learned memories. My friend and longtime collaborator at Harvard Medical School, Dr. Robert Stickgold, conducted a clever study with wide-reaching implications. He had a total of 133 undergraduates learn a visual memory task through repetition. Participants then returned to his laboratory and were tested to see how much they had retained. Some subjects returned the next day after a full night of sleep. Others returned two days later after two full nights of sleep, and still others after three days with three nights of sleep in between.

As you would predict by now, a night of sleep strengthened the newly learned memories, boosting their retention. Additionally, the more nights of sleep participants had before they were tested, the better their memory was. All except another sub-group of participants. Like the subjects in the third group, these participants learned the task on the first day, and learned it just as well. They were then tested three nights later, just like the third group above. The difference was that they were deprived of sleep the first night after learning and were not tested the following day. Instead, Stickgold gave them two full recovery nights of sleep before testing them. They showed absolutely no evidence of a memory consolidation improvement. In other words, if you don't sleep the very first night after learning, you lose the chance to consolidate those memories, even if you get lots of "catch-up" sleep thereafter. In terms of memory, then, sleep is not like the bank. You cannot accumulate a debt and hope to pay it off at a later point in time. Sleep for memory consolidation is an all-or-nothing event. It is a concerning result in our 24/7, hurry-up, don't-wait society. I feel another op-ed coming on . . .

SLEEP AND ALZHEIMER'S DISEASE

The two most feared diseases throughout developed nations are dementia and cancer. Both are related to inadequate sleep. We will address the latter in the next chapter regarding sleep deprivation and the body. Regarding the former, which centers on the brain, a lack of sleep is becoming recognized as a lifestyle factor contributing to your risk of developing Alzheimer's disease.

The condition, originally identified in 1901 by German physician Dr. Aloysius Alzheimer, has become one of the largest public health and economic challenges of the twenty-first century. More than 40 million people suffer from the debilitating disease. That number has accelerated as the human life span has stretched, but also, importantly, as total sleep time has decreased. One in ten adults over the age of sixty-five now suffers from Alzheimer's disease. Without advances in diagnosis, prevention, and therapeutics, the escalation will continue.

Sleep represents a new candidate for hope on all three of these fronts: diagnosis, prevention, and therapeutics. Before discussing why, let me first describe how sleep disruption and Alzheimer's disease are causally linked.

As we learned in chapter 5, sleep quality—especially that of deep NREM sleep—deteriorates as we age. This is linked to a decline in memory. However, if you assess a patient with Alzheimer's disease, the disruption of deep sleep is far more exaggerated. More telling, perhaps, is the fact that sleep disturbance precedes the onset of Alzheimer's disease by several years, suggesting that it may be an early-warning sign of the condition, or even a contributor to it. Following diagnosis, the magnitude of sleep disruption will then progress in unison with the symptom severity of the Alzheimer's patient, further suggesting a link between the two. Making matters worse, over 60 percent of patients with Alzheimer's disease have at least one clinical sleep disorder. Insomnia is especially common, as caregivers of a loved one with Alzheimer's disease will know all too well.

It was not until relatively recently, however, that the association between disturbed sleep and Alzheimer's disease was realized to be more than just an association. While much remains to be understood, we now recognize that sleep disruption and Alzheimer's disease interact in a self-fulfilling, negative spiral that can initiate and/or accelerate the condition.

Alzheimer's disease is associated with the buildup of a toxic form of protein called beta-amyloid, which aggregates in sticky clumps, or plaques, within the brain. Amyloid plaques are poisonous to neurons, killing the surrounding brain cells. What is strange, however, is that amyloid plaques only affect some parts of the brain and not others, the reasons for which remain unclear.

What struck me about this unexplained pattern was the location in the brain where amyloid accumulates early in the course of Alzheimer's disease, and most severely in the late stages of the condition. That area is the middle part of the frontal lobe—which, as you will remember, is the same brain region essential for the electrical generation of deep NREM sleep in healthy young individuals. At that time, we did not understand if or why Alzheimer's disease caused sleep disruption, but simply knew that they always co-occurred. I wondered whether the reason patients with Alzheimer's disease have such impaired deep NREM sleep was, in part, because the disease erodes the very region of the brain that normally generates this key stage of slumber.

I joined forces with Dr. William Jagust, a leading authority on Alzheimer's disease, at the University of California, Berkeley. Together, our research teams set about testing this hypothesis. Several years later, having assessed the sleep of many older adults with varying degrees of amyloid buildup in the brain that we quantified with a special type of PET scan, we arrived at the answer. The more amyloid deposits there were in the middle regions of the frontal lobe, the more impaired the deep-sleep quality was in that older individual. And it was not just a general loss of deep sleep, which is common as we get older, but the very deepest of the powerful slow brainwaves of NREM sleep that the disease was ruthlessly eroding. This distinction was important, since it meant that the sleep impairment caused by amyloid buildup in the brain was more than just "normal aging." It was unique—a departure from what is otherwise the signature of sleep decline as we get older.

We are now examining whether this very particular "dent" in sleeping brainwave activity represents an early identifier of those who are at greatest risk of developing Alzheimer's disease, years in advance. If sleep does prove to be an early diagnostic measure—especially one that is relatively cheap, noninvasive, and can be easily obtained in a large number of individuals, unlike costly MRI or PET scans—then early intervention becomes possible.

Building on these findings, our recent work has added a key piece in the jigsaw puzzle of Alzheimer's disease. We have discovered a new pathway through which amyloid plaques may contribute to memory decline later in life: something that has been largely missing in our understanding of how Alzheimer's disease works. I mentioned that the toxic amyloid deposits only accumulate in some parts of the brain and not others. Despite Alzheimer's disease being typified by memory loss, the hippocampus—that key memory reservoir in the brain—is mysteriously unaffected by amyloid protein. This question has so far baffled scientists: How does amyloid cause memory loss in Alzheimer's disease patients when amyloid itself does not affect memory areas of the brain? While other aspects of the disease may be at play, it seemed plausible to me that there was a missing intermediary factor—one that was transacting the influence of amyloid in one part of the brain on memory, which depended on a different region of the brain. Was sleep disruption the missing factor?

To test this theory, we had elderly patients with varying levels of amyloid—low to high—in their brains learn a list of new facts in the evening. The next morning, after recording their sleep in the laboratory that night, we tested them to see how effective their sleep had been at cementing and thus holding on to those new memories. We discovered a chain-reaction effect. Those individuals with the highest levels of amyloid deposits in the frontal regions of the brain had the most severe loss of deep sleep and, as a knock-on consequence, failed to successfully consolidate those new memories. Overnight *forgetting*, rather than remembering, had taken place. The disruption of deep NREM sleep was therefore a hidden middleman brokering the bad deal between amyloid and memory impairment in Alzheimer's disease. A missing link.

These findings, however, were only half of the story, and admittedly the less important half. Our work had shown that the amyloid plaques of Alzheimer's disease may be associated with the loss of deep sleep, but does it work both ways? Can a lack of sleep actually cause amyloid to build up in your brain to begin with? If so, insufficient sleep across an individual's life would significantly raise their risk of developing Alzheimer's disease.

Around the same time that we were conducting our studies, Dr. Maiken Nedergaard at the University of Rochester made one of the most spectacular discoveries in the field of sleep research in recent decades. Working with mice, Nedergaard found that a kind of sewage network called the glymphatic system exists within the brain. Its name is derived from the body's equivalent lymphatic system, but it's composed of cells called glia (from the Greek root word for "glue").

Glial cells are distributed throughout your entire brain, situated side by side with the neurons that generate the electrical impulses of your brain. Just as the lymphatic system drains contaminants from your body, the glymphatic system collects and removes dangerous metabolic contaminants generated by the hard work performed by neurons in your brain, rather like a support team surrounding an elite athlete.

Although the glymphatic system—the support team—is somewhat active during the day, Nedergaard and her team discovered that it is during sleep that this neural sanitization work kicks into high gear. Associated with the pulsing rhythm of deep NREM sleep comes a ten- to twentyfold increase in effluent expulsion from the brain. In what can be described as a nighttime power cleanse, the purifying work of the glymphatic system is accomplished by cerebrospinal fluid that bathes the brain.

Nedergaard made a second astonishing discovery, which explained why the cerebrospinal fluid is so effective in flushing out metabolic debris at night. The glial cells of the brain were shrinking in size by up to 60 percent during NREM sleep, enlarging the space around the neurons and allowing the cerebrospinal fluid to proficiently clean out the metabolic refuse left by the day's neural activity. Think of the buildings of a large metropolitan city physically shrinking at night, allowing municipal cleaning crews easy access to pick up garbage strewn in the streets, followed by a good pressure-jet treatment of every nook and cranny. When we wake each morning, our brains can once again function efficiently thanks to this deep cleansing.

So what does this have to do with Alzheimer's disease? One piece of toxic debris evacuated by the glymphatic system during sleep is amyloid protein—the poisonous element associated with Alzheimer's disease. Other dangerous metabolic waste elements that have links to Alzheimer's disease are also removed by the cleaning process during sleep, including a protein called tau, as well as stress molecules produced by neurons when they combust energy and oxygen during the day. Should you experimentally prevent a mouse from getting NREM sleep, keeping it awake instead, there is an immediate increase in amyloid deposits within the brain. Without sleep, an escalation of poisonous Alzheimer's-related protein accumulated in the brains of the mice, together with several other toxic metabolites. Phrased differently, and perhaps more simply, wakefulness is low-level brain damage, while sleep is neurological sanitation.

Nedergaard's findings completed the circle of knowledge that our findings had left unanswered. Inadequate sleep and the pathology of Alzheimer's disease interact in a vicious cycle. Without sufficient sleep, amyloid plaques build up in the brain, especially in deep-sleep-generating regions, attacking and degrading them. The loss of deep NREM sleep caused by this assault therefore lessens the ability to remove amyloid from the brain at night, resulting in greater amyloid deposition. More amyloid, less deep sleep, less deep sleep, more amyloid, and so on and so forth.

From this cascade comes a prediction: getting too little sleep across the adult life span will significantly raise your risk of developing Alzheimer's disease. Precisely this relationship has now been reported in numerous epidemiological studies, including those individuals suffering from sleep disorders such as insomnia and sleep apnea.* Par-

*A. S. Lim et al., "Sleep Fragmentation and the Risk of Incident Alzheimer's Disease and Cognitive Decline in Older Persons," *Sleep* 36 (2013): 1027–32; A. S. Lim et al., "Modification of the relationship of the apolipoprotein E epsilon4 allele to the risk of Alzheimer's disease and neurofibrillary tangle density by sleep," *JAMA Neurology* 70 (2013): 1544–51; R.

enthetically, and unscientifically, I have always found it curious that Margaret Thatcher and Ronald Reagan—two heads of state that were very vocal, if not proud, about sleeping only four to five hours a night—both went on to develop the ruthless disease. The current US president, Donald Trump—also a vociferous proclaimer of sleeping just a few hours each night—may want to take note.

A more radical and converse prediction that emerges from these findings is that, by improving someone's sleep, we should be able to reduce their risk of developing Alzheimer's disease—or at least delay its onset. Tentative support has emerged from clinical studies in which middle- and older-age adults have had their sleep disorders successfully treated. As a consequence, their rate of cognitive decline slowed significantly, and further delayed the onset of Alzheimer's disease by five to ten years.*

My own research group is now trying to develop a number of viable methods for artificially increasing deep NREM sleep that could restore some degree of the memory consolidation function that is absent in older individuals with high amounts of amyloid in the brain. If we can find a method that is cost effective and can be scaled up to the population level for repeat use, my goal is prevention. Can we begin supplementing the declining deep sleep of vulnerable members of society during midlife, many decades before the tipping point of Alzheimer's disease is reached, aiming to avert dementia risk later in life? It is an admittedly lofty ambition, and some would argue a moon shot research goal. But it is worth recalling that we already use this conceptual approach in medicine in the form of prescribing statins to higher-risk individuals in their forties and fifties to help prevent cardiovascular disease, rather than having to treat it decades later.

S. Osorio et al., "Greater risk of Alzheimer's disease in older adults with insomnia," *Journal of the American Geriatric Society* 59 (2011): 559–62; and K. Yaffe et al., "Sleep-disordered breathing, hypoxia, and risk of mild cognitive impairment and dementia in older women," *JAMA* 306 (2011): 613–19.

*S. Ancoli-Israel et al., "Cognitive effects of treating obstructive sleep apnea in Alzheimer's disease: a randomized controlled study," *Journal of the American Geriatric Society* 56 (2008): 2076–81; and W.d.S. Moraes et al., "The effect of donepezil on sleep and REM sleep EEG in patients with Alzheimer's disease: a double-blind placebo-controlled study," *Sleep* 29 (2006): 199–205.

Insufficient sleep is only one among several risk factors associated with Alzheimer's disease. Sleep alone will not be the magic bullet that eradicates dementia. Nevertheless, prioritizing sleep across the life span may be a significant factor for lowering Alzheimer's disease risk.

CHAPTER 8

Cancer, Heart Attacks, and a Shorter Life

Sleep Deprivation and the Body

I was once fond of saying, "Sleep is the third pillar of good health, alongside diet and exercise." I have changed my tune. Sleep is more than a pillar; it is the foundation on which the other two health bastions sit. Take away the bedrock of sleep, or weaken it just a little, and careful eating or physical exercise become less than effective, as we shall see.

Yet the insidious impact of sleep loss on health runs much deeper. Every major system, tissue, and organ of your body suffers when sleep becomes short. No aspect of your health can retreat at the sign of sleep loss and escape unharmed. Like water from a burst pipe in your home, the effects of sleep deprivation will seep into every nook and cranny of biology, down into your cells, even altering your most fundamental self—your DNA.

Widening the lens of focus, there are more than twenty large-scale epidemiological studies that have tracked millions of people over many decades, all of which report the same clear relationship: the shorter your sleep, the shorter your life. The leading causes of disease and death in developed nations—diseases that are crippling health-care systems, such as heart disease, obesity, dementia, diabetes, and cancer—all have recognized links to insufficient or disrupted sleep.

This chapter describes varied ways in which insufficient sleep proves ruinous to all the major physiological systems of the human body: cardiovascular, metabolic, immune, reproductive.

SLEEP LOSS AND THE CARDIOVASCULAR SYSTEM

Unhealthy sleep, unhealthy heart. Simple and true. Take the results of a 2011 study that tracked 474,684 men and women of varied ages, races, and ethnicities across eight different countries. Short sleep was associated with more than a 45 percent greater risk of fatal and nonfatal coronary heart disease within seven to twenty-five years from the start of the study. A similar relationship was observed in a Japanese study of 2,282 male workers. Over a fourteen-year period, sleeping less than six hours a night was associated with more than a three times greater risk of suffering a cardiovascular or coronary event, such as sudden cardiac death. This was relative to those sleeping 7 to 7.9 hours a night. I should note that in many of these studies, the relationship between short sleep and heart failure remains strong even after controlling for other known cardiac risk factors, such as smoking, physical activity, and body mass. A lack of sleep more than accomplishes its own, independent attack on the heart.

As we approach midlife, and our body begins to deteriorate and health resilience starts its decline, the impact of insufficient sleep on the cardiovascular system escalates. Adults forty-five years or older who sleep fewer than six hours a night are 200 percent more likely to have a heart attack or stroke during their lifetime, as compared with those sleeping seven to eight hours a night. This finding impresses how important it is to prioritize sleep in midlife—which is unfortunately the time when family and professional circumstances encourage us to do the exact opposite.

Part of the reason the heart suffers so dramatically under the weight of sleep deprivation concerns blood pressure. Have a quick look at your right forearm and pick out some veins. If you wrap your left hand around that forearm, just below the elbow, and grip it, like a tourniquet, you will see those vessels start to balloon. A little alarming, isn't it? The ease with which just a little sleep loss can pump up pressure in the veins of your entire body, stretching and distressing the vessel walls, is equally alarming. High blood pressure is so common nowadays that we forget the deathly toll it inflicts. This year alone, hypertension will steal more than 7 million people's lives by way of cardiac failure, ischemic heart disease, stroke, or kidney failure. Deficient sleep is responsible for many of these lost fathers, mothers, grandparents, and beloved friends.

As with other consequences of sleep loss we've encountered, you don't need a full night of total sleep deprivation to inflict a measurable impact on your cardiovascular system. One night of modest sleep reduction—even just one or two hours—will promptly speed the contracting rate of a person's heart, hour upon hour, and significantly increase the systolic blood pressure within their vasculature.* You will find no solace in the fact that these experiments were conducted in young, fit individuals, all of whom started out with an otherwise healthy cardiovascular system just hours before. Such physical fitness proves no match for a short night of sleep; it affords no resistance.

Beyond accelerating your heart rate and increasing your blood pressure, a lack of sleep further erodes the fabric of those strained blood vessels, especially those that feed the heart itself, called the coronary arteries. These corridors of life need to be clean and open wide to supply your heart with blood at all times. Narrow or block those passageways, and your heart can suffer a comprehensive and often fatal attack caused by blood oxygen starvation, colloquially known as a "massive coronary."

One cause of a coronary artery blockage is atherosclerosis, or the furring up of those heart corridors with hardened plaques that contain calcium deposits. Researchers at the University of Chicago studied almost five hundred healthy midlife adults, none of whom had any existing heart disease or signs of atherosclerosis. They tracked the health of the coronary arteries of these participants for a number of years, all the while assessing their sleep. If you were one of the individuals who were obtaining just five to six hours each night or less, you were 200 to 300 percent more likely to suffer calcification of your coronary arteries over the next five years, relative to those individuals sleeping seven to eight hours. The deficient sleep of those individuals was associated with a closing off of the critical passageways that should otherwise be

*O. Tochikubo, A. Ikeda, E. Miyajima, and M. Ishii, "Effects of insufficient sleep on blood pressure monitored by a new multibiomedical recorder," *Hypertension* 27, no. 6 (1996): 1318–24.

wide open and feeding the heart with blood, starving it and significantly increasing the risk of a coronary heart attack.

Although the mechanisms by which sleep deprivation degrades cardiovascular health are numerous, they all appear to cluster around a common culprit, called the sympathetic nervous system. Abandon any thoughts of love or serene compassion based on the misguiding name. The sympathetic nervous system is resolutely activating, inciting, even agitating. If needed, it will mobilize the evolutionarily ancient fight-or-flight stress response within the body, comprehensively and in a matter of seconds. Like an accomplished general in command of a vast military, the sympathetic nervous system can muster activity in a vast assortment of the body's physiological divisions—from respiration, immune function, and stress chemicals to blood pressure and heart rate.

An acute stress response from the sympathetic nervous system, which is normally only deployed for short periods of time lasting minutes to hours, can be highly adaptive under conditions of credible threat, such as the potential of real physical attack. Survival is the goal, and these responses promote immediate action to accomplish just that. But leave that system stuck in the "on" position for long durations of time, and sympathetic activation becomes deeply maladaptive. In fact, it is a killer.

Many experiments over the past half century have investigated the impact of deficient sleep on the human body and typically observed an overactive sympathetic nervous system. For as long as the state of insufficient sleep lasts, and for some time thereafter, the body remains stuck in some degree of a fight-or-flight state. It can last for years in those with an untreated sleep disorder, excessive work hours that limit sleep or its quality, or the simple neglect of sleep by an individual. Like a car engine that is revved to a shrieking extreme for sustained periods of time, your sympathetic nervous system is floored into perpetual overdrive by a lack of sleep. The consequential strain that is placed on your body by the persistent force of sympathetic activation will leak out in all manner of health issues, just like the failed pistons, gaskets, seals, and gnashing gears of an abused car engine.

Through this central pathway of an overactive sympathetic nervous system, together with increases in a stress-related chemical called cortisol, sleep deprivation triggers a domino effect that will spread like a wave of health damage throughout your body. It starts with removing a default resting brake that normally prevents your heart from accelerating in its rate of contraction. Once this brake is released, you will experience sustained speeds of cardiac beating.

As your sleep-deprived heart beats faster, the volumetric rate of blood pumped through your vasculature increases, and with that comes the hypertensive state of your blood pressure. Occurring at the same time is a chronic increase in a stress hormone called cortisol, which is triggered by the overactive sympathetic nervous system. One undesirable consequence of the sustained deluge of cortisol is the constriction of those blood vessels, triggering an even greater increase in blood pressure.

Making matters worse, growth hormone—a great healer of the body—which normally surges at night, is shut off by the state of sleep deprivation. Without growth hormone to replenish the lining of your blood vessels, called the endothelium, they will be slowly shorn and stripped of their integrity. Adding insult to real injury, the hypertensive strain that sleep deprivation places on your vasculature means that you can no longer repair those fracturing vessels effectively. The damaged and weakened state of vascular plumbing throughout your body now becomes systemically more prone to atherosclerosis (arteries furring up). Vessels will rupture. It is a powder keg of factors, with heart attack and stroke being the most common casualties in the explosive aftermath.

Compare this cascade of harm to the healing benefits that a full night of sleep normally lavishes on the cardiovascular system. During deep NREM sleep specifically, the brain communicates a calming signal to the fight-or-flight sympathetic branch of the body's nervous system, and does so for long durations of the night. As a result, deep sleep prevents an escalation of this physiological stress that is synonymous with increased blood pressure, heart attack, heart failure, and stroke. This includes a calming effect on the contracting speed of your heart. Think of your deep NREM sleep as a natural form of nighttime blood-pressure management—one that averts hypertension and stroke.

When communicating science to the general public in lectures or writing, I'm always wary of bombarding an audience with never-ending mortality and morbidity statistics, lest they themselves lose the will to live in front of me. It is hard not to do so with such compelling masses of studies in the field of sleep deprivation. Often, however, a single astonishing result is all that people need to apprehend the point. For cardiovascular health, I believe that finding comes from a "global experiment" in which 1.5 billion people are forced to reduce their sleep by one hour or less for a single night each year. It is very likely that you have been part of this experiment, otherwise known as daylight savings time.

In the Northern Hemisphere, the switch to daylight savings time in March results in most people losing an hour of sleep opportunity. Should you tabulate millions of daily hospital records, as researchers have done, you discover that this seemingly trivial sleep reduction comes with a frightening spike in heart attacks the following day. Impressively, it works both ways. In the autumn within the Northern Hemisphere, when the clocks move forward and we gain an hour of sleep opportunity time, rates of heart attacks plummet the day after. A similar rise-and-fall relationship can be seen with the number of traffic accidents, proving that the brain, by way of attention lapses and microsleeps, is just as sensitive as the heart to very small perturbations of sleep.

SLEEP LOSS AND METABOLISM: DIABETES AND WEIGHT GAIN

The less you sleep, the more you are likely to eat. In addition, your body becomes unable to manage those calories effectively, especially the concentrations of sugar in your blood. In these two ways, sleeping less than seven or eight hours a night will increase your probability of gaining weight, being overweight, or being obese, and significantly increases your likelihood of developing type 2 diabetes.

The global health cost of diabetes is $375 billion a year. That of obesity is more than $2 trillion. Yet for the under-slept individual, the cost to health, quality of life, and a hastened arrival of death are more meaningful. Precisely how a lack of sleep sets you on a path toward diabetes and leads to obesity is now well understood and incontrovertible.

DIABETES

Sugar is a dangerous thing. In your diet, yes, but here I'm referring to that which is currently circulating in your bloodstream. Excessively high levels of blood sugar, or glucose, over weeks or years inflicts a surprising harm to the tissues and organs of your body, worsens your health, and shortens your life span. Eye disease that can end in blindness, nerve disease that commonly results in amputations, and kidney failure necessitating dialysis or transplant are all consequences of prolonged high blood sugar, as are hypertension and heart disease. But it is the condition of type 2 diabetes that is most commonly and immediately related to unregulated blood sugar.

In a healthy individual, the hormone insulin will trigger the cells of your body to swiftly absorb glucose from the bloodstream should it increase, as happens after eating a meal. Instructed by insulin, the cells of your body will open special channels on their surface that operate like wonderfully efficient roadside drains at the height of a downpour. They have no problem dealing with the deluge of glucose coursing down the transit arteries, averting what could otherwise be a dangerous flood of sugar in the bloodstream.

If the cells of your body stop responding to insulin, however, they cannot efficiently absorb glucose from the blood. Similar to roadside drains that become blocked or erroneously closed shut, the rising swell of blood sugar cannot be brought back down to safe levels. At this point, the body has transitioned into a hyperglycemic state. Should this condition persist, and the cells of your body remain intolerant to dealing with the high levels of glucose, you will transition into a pre-diabetic state and, ultimately, develop full-blown type 2 diabetes.

Early-warning signs of a link between sleep loss and abnormal blood sugar emerged in a series of large epidemiological studies spanning several continents. Independent of one another, the research groups found far higher rates of type 2 diabetes among individuals that reported sleeping less than six hours a night routinely. The association remained significant even when adjusting for other contributing factors, such as body weight, alcohol, smoking, age, gender, race, and caffeine use. Powerful as these studies are, though, they do not inform the direction of causality. Does the state of diabetes impair your sleep, or does short sleep impair your body's ability to regulate blood sugar, thereby causing diabetes?

To answer this question, scientists had to conduct carefully controlled experiments with healthy adults who had no existing signs of diabetes or issues with blood sugar. In the first of these studies, participants were limited to sleeping four hours a night for just six nights. By the end of that week, these (formerly healthy) participants were 40 percent less effective at absorbing a standard dose of glucose, compared to when they were fully rested.

To give you a sense of what that means, if the researchers showed those blood sugar readings to an unwitting family doctor, the GP would immediately classify that individual as being pre-diabetic. They would start a rapid intervention program to prevent the development of irreversible type 2 diabetes. Numerous scientific laboratories around the world have replicated this alarming effect of short sleep, some with even less aggressive reductions in sleep amount.

How does a lack of sleep hijack the body's effective control of blood sugar? Was it a blockade of insulin release, removing the essential instruction for cells to absorb glucose? Or had the cells themselves become unresponsive to an otherwise normal and present message of insulin?

As we have discovered, both are true, though the most compelling evidence indicates the latter. By taking small tissue samples, or biopsies, from participants at the end of the above experiments, we can examine how the cells of the body are operating. After participants had been restricted to four to five hours of sleep for a week, the cells of these tired individuals had become far less receptive to insulin. In this sleep-deprived state, the cells were stubbornly resisting the message from insulin and refusing to open up their surface channels. The roadside drains were effectively closed shut, leading to a rising tide of glucose in the bloodstream and a pre-diabetic state of hyperglycemia.

While many in the general public understand that diabetes is serious, they may not appreciate the true burden. Beyond the average treatment cost of more than $85,000 per patient (which contributes to higher medical insurance premiums), diabetes lops ten years off an individual's life expectancy. Chronic sleep deprivation is now recognized as one of the major contributors to the escalation of type 2 diabetes throughout first-world countries. It's a preventable contribution.

WEIGHT GAIN AND OBESITY

When your sleep becomes short, you will gain weight. Multiple forces conspire to expand your waistline. The first concerns two hormones controlling appetite: leptin and ghrelin.* Leptin signals a sense of feeling full. When circulating levels of leptin are high, your appetite is blunted and you don't feel like eating. Ghrelin, in contrast, triggers a strong sensation of hunger. When ghrelin levels increase, so, too, does your desire to eat. An imbalance of either one of these hormones can trigger increased eating and thus body weight. Perturb both in the wrong direction, and weight gain is more than probable.

Over the past thirty years, my colleague Dr. Eve Van Cauter at the University of Chicago has tirelessly conducted research on the link between sleep and appetite that is as brilliant as it is impactful. Rather than depriving individuals of a full night of sleep, Van Cauter has taken a more relevant approach. She recognized that many individuals in industrialized societies sleep less than five to six hours a night during the week. So in a first series of studies of healthy young adults of perfectly normal weight, she began to investigate whether one week of this societally typical short sleep was enough to disrupt levels of either leptin or ghrelin or both.

If you are a participant in one of Van Cauter's studies, it feels rather more like a one-week stay at a hotel. You will get your own room, bed, clean sheets, a television, Internet access, etc.—everything except free tea and coffee, since no caffeine is allowed. In one arm of the experiment, you will be given an eight-and-a-half-hour sleep opportunity each night for five nights, recorded with electrodes placed on your head. In the other arm of the study, you are only allowed four to five hours of sleep for five nights, also measured with electrode recordings. In both study arms, you will receive exactly the same amount and type of food, and your degree of physical activity is also held constant. Each day, your sense of hunger and food intake are monitored, as are your circulating levels of leptin and ghrelin.

Using precisely this experimental design in a group of healthy, lean

*While leptin and ghrelin may sound like the names of two hobbits, the former is derived from the Greek term *leptos*, meaning slender, while the latter comes from *ghre*, the Proto-Indo-European term for growth.

participants, Van Cauter discovered that individuals were far more ravenous when sleeping four to five hours a night. This despite being given the same amount of food and being similarly active, which kept the hunger levels of these same individuals under calm control when they were getting eight or more hours of sleep. The strong rise of hunger pangs and increased reported appetite occurred rapidly, by just the second day of short sleeping.

At fault were the two characters, leptin and ghrelin. Inadequate sleep decreased concentrations of the satiety-signaling hormone leptin and increased levels of the hunger-instigating hormone ghrelin. It was a classic case of physiological double jeopardy: participants were being punished twice for the same offense of short sleeping: once by having the "I'm full" signal removed from their system, and once by gaining the "I'm still hungry" feeling being amplified. As a result, participants just didn't feel satisfied by food when they were short sleeping.

From a metabolic perspective, the sleep-restricted participants had lost their hunger control. By limiting these individuals to what some in our society would think of as a "sufficient" amount of sleep (five hours a night), Van Cauter had caused a profound imbalance in the scales of hormonal food desire. By muting the chemical message that says "stop eating" (leptin), yet increasing the hormonal voice that shouts "please, keep eating" (ghrelin), your appetite remains unsatisfied when your sleep is anything less than plentiful, even after a kingly meal. As Van Cauter has elegantly described to me, a sleep-deprived body will cry famine in the midst of plenty.

But feeling hungry and actually eating more are not the same thing. Do you actually eat more when sleeping less? Does your waistline really swell as a consequence of that rise in appetite?

With another landmark study, Van Cauter proved this to be the case. Participants in this experiment again underwent two different conditions, acting as their own baseline control: four nights of eight and a half hours' time in bed, and four nights of four and a half hours' time in bed. Each day, participants were limited to the same level of physical activity under both conditions. Each day, they were given free access to food, and the researchers meticulously counted the difference in calorie consumption between the two experimental manipulations.

When short sleeping, the very same individuals ate 300 calories more each day—or well over 1,000 calories before the end of the experiment—compared to when they were routinely getting a full night of sleep. Similar changes occur if you give people five to six hours of sleep over a ten-day period. Scale that up to a working year, and assuming one month of vacation in which sleep miraculously becomes abundant, and you will still have consumed more than 70,000 extra calories. Based on caloric estimates, that could cause 10 to 15 pounds of weight gain a year, each and every year.

Van Cauter's next experiment was the most surprising (and devilish) of all. Fit, healthy individuals went through the same two different conditions as before: four nights of eight and a half hours' time in bed, and four nights of four and a half hours' time in bed. However, on the last day of each of the experimental conditions, something different happened. Participants were offered an additional food buffet stretched across a four-hour period. Set out in front of them was an assortment of foods, from meats, vegetables, bread, potatoes, and salad to fruit and ice cream. Set to one side, however, was access to a bonus snack bar filled with cookies, chocolate bars, chips, and pretzels. Participants could eat as much as they wanted in the four-hour period, with the buffet even being replenished halfway through. Importantly, the subjects ate alone, limiting social or stigmatizing influences that could alter their natural eating urges.

Following the buffet, Van Cauter and her team once again quantified what participants ate, and how much they ate. Despite eating almost 2,000 calories during the buffet lunch, sleep-deprived participants dove into the snack bar. They consumed an *additional* 330 calories of snack foods after the full meal, compared to when they were getting plenty of sleep each night.

Of relevance to this behavior is a recent discovery that sleep loss increases levels of circulating endocannabinoids, which, as you may have guessed from the name, are chemicals produced by the body that are very similar to the drug cannabis. Like marijuana use, these chemicals stimulate appetite and increase your desire to snack, otherwise known as having the munchies.

Combine this increase in endocannabinoids with alterations in leptin and ghrelin caused by sleep deprivation and you have a potent brew of chemical messages all driving you in one direction: overeating.

Some argue that we eat more when we are sleep-deprived because we burn extra calories when we stay awake. Sadly, this is not true. In the sleep-restriction experiments described above, there are no differences in caloric expenditure between the two conditions. Take it to the extreme by sleep-depriving an individual for twenty-four hours straight and they may only burn an extra 147 calories, relative to a twenty-four-hour period containing a full eight hours of sleep. Sleep, it turns out, is an intensely metabolically active state for brain and body alike. For this reason, theories proposing that we sleep to conserve large amounts of energy are no longer entertained. The limited caloric savings are insufficient to outweigh the survival dangers and disadvantages associated with falling asleep.

More importantly, the extra calories that you eat when sleep-deprived can outweigh any nominal extra energy you burn while remaining awake. Making matters worse, the less an individual sleeps, the less energy he or she feels they have, and the more sedentary and less willing to exercise they are in real-world settings. Inadequate sleep is the perfect recipe for obesity: greater calorie intake, lower calorie expenditure.

Weight gain caused by short sleep is not just a matter of eating more, but also a change in *what* you binge eat. Looking across the different studies, Van Cauter noticed that cravings for sweets (e.g., cookies, chocolate, and ice cream), heavy-hitting carbohydrate-rich foods (e.g., bread and pasta), and salty snacks (e.g., potato chips and pretzels) all increased by 30 to 40 percent when sleep was reduced by several hours each night. Less affected were protein-rich foods (e.g., meat and fish), dairy items (such as yogurt and cheese), and fatty foods, showing a 10 to 15 percent increase in preference by the sleepy participants.

Why is it that we lust after quick-fix sugars and complex carbohydrates when sleep-deprived? My research team and I decided to conduct a study in which we scanned people's brains while they were viewing and choosing food items, and then rated how much they desired each one. We hypothesized that changes within the brain may help explain this unhealthy shift in food preference caused by a lack of sleep. Was there a breakdown in impulse-control regions that normally keep our basic hedonic food desires in check, making us reach for doughnuts or pizza rather than whole grains and leafy greens?

Healthy, average-weight participants performed the experiment twice: once when they had had a full night of sleep, and once after they had been sleep-deprived for a night. In each of the two conditions they viewed eighty similar food images, ranging from fruits and vegetables, such as strawberries, apples, and carrots, to high-calorie items, such as ice cream, pasta, and doughnuts. To ensure that participants were making choices that reflected their true cravings rather than simply choosing items that they thought would be the right or most appropriate choice, we forced an incentive: after they came out of the MRI machine, we gave them a serving of the food they told us they most craved during the task, and politely asked them to eat it!

Comparing the patterns of brain activity between the two conditions within the same individual, we discovered that supervisory regions in the prefrontal cortex required for thoughtful judgments and controlled decisions had been silenced in their activity by a lack of sleep. In contrast, the more primal deep-brain structures that drive motivations and desire were amplified in response to the food images. This shift to a more primitive pattern of brain activity without deliberative control came with a change in the participants' food choices. High-calorie foods became significantly more desirable in the eyes of the participants when sleep-deprived. When we tallied up the extra food items that participants wanted when they were sleep-deprived, it amounted to an extra 600 calories.

The encouraging news is that getting enough sleep will help you control body weight. We found that a full night of sleep repairs the communication pathway between deep-brain areas that unleash hedonic desires and higher-order brain regions whose job it is to rein in these cravings. Ample sleep can therefore restore a system of impulse control within your brain, putting the appropriate brakes on potentially excessive eating.

South of the brain, we are also discovering that plentiful sleep makes your gut happier. Sleep's role in redressing the balance of the body's nervous system, especially its calming of the fight-or-flight sympathetic branch, improves the bacterial community known as your microbiome, which is located in your gut (also known as the enteric nervous system). As we learned about earlier, when you do not get enough sleep, and the body's stress-related, fight-or-flight nervous system is revved up, this triggers an excess of circulating cortisol that cultivates "bad bacteria" to fester throughout your microbiome. As a result, insufficient sleep will prevent the meaningful absorption of all food nutrients and cause gastrointestinal problems.*

Of course, the obesity epidemic that has engulfed large portions of the world is not caused by lack of sleep alone. The rise in consumption of processed foods, an increase in serving sizes, and the more sedentary nature of human beings are all triggers. However, these changes are insufficient to explain the dramatic escalation of obesity. Other factors must be at play.

Based on evidence gathered over the past three decades, the epidemic of insufficient sleep is very likely a key contributor to the epidemic of obesity. Epidemiological studies have established that people who sleep less are the same individuals who are more likely to be overweight or obese. Indeed, if you simply plot the reduction in sleep time (dotted line) over the past fifty years on the same graph as the rise in obesity rates across the same time period (solid line), shown in Figure 13, the data infer, but do not prove, this relationship.

Figure 13: Sleep Loss and Obesity

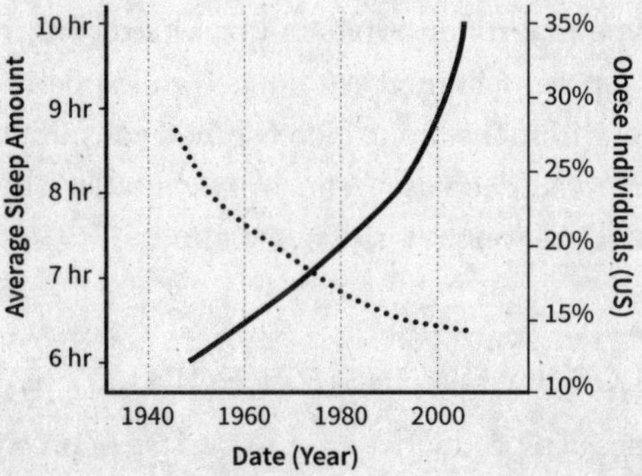

*I suspect we'll discover a two-way relationship wherein sleep not only affects the microbiome, but the microbiome can communicate with and alter sleep through numerous different biological channels.

We are now observing these effects very early in life. Three-year-olds sleeping just ten and a half hours or less have a 45 percent increased risk of being obese by age seven than those who get twelve hours of sleep a night. To set our children on a pathway of ill health this early in life by way of sleep neglect is a travesty.

A final comment on trying to lose weight: let's say that you choose to go on a strict, low-calorie diet for two weeks in the hopes of losing fat and looking more trim and toned as a consequence. That's precisely what researchers did to a group of overweight men and women who stayed in a medical center for an entire fortnight. However, one group of individuals were given just five and a half hours' time in bed, while the other group were offered eight and a half hours' time in bed.

Although weight loss occurred under both conditions, the *type* of weight loss came from very different sources. When given just five and a half hours of sleep opportunity, more than 70 percent of the pounds lost came from lean body mass—muscle, not fat. Switch to the group offered eight and a half hours' time in bed each night and a far more desirable outcome was observed, with well over 50 percent of weight loss coming from fat while preserving muscle. When you are not getting enough sleep, the body becomes especially stingy about giving up fat. Instead, muscle mass is depleted while fat is retained. Lean and toned is unlikely to be the outcome of dieting when you are cutting sleep short. The latter is counterproductive of the former.

The upshot of all this work can be summarized as follows: short sleep (of the type that many adults in first-world countries commonly and routinely report) will increase hunger and appetite, compromise impulse control within the brain, increase food consumption (especially of high-calorie foods), decrease feelings of food satisfaction after eating, and prevent effective weight loss when dieting.

SLEEP LOSS AND THE REPRODUCTIVE SYSTEM

If you have hopes of reproductive success, fitness, or prowess, you would do well to get a full night's sleep every night. Charles Darwin would, I'm sure, cleave easily to this advice, had he reviewed the evidence I now present.

Take a group of lean, healthy young males in their mid-twenties and limit them to five hours of sleep for one week, as a research group did at the University of Chicago. Sample the hormone levels circulating in the blood of these tired participants and you will find a marked drop in testosterone relative to their own baseline levels of testosterone when fully rested. The size of the hormonal blunting effect is so large that it effectively "ages" a man by ten to fifteen years in terms of testosterone virility. The experimental results support the finding that men suffering from sleep disorders, especially sleep apnea associated with snoring, have significantly lower levels of testosterone than those of similar age and backgrounds but who do not suffer from a sleep condition.

Uttering the results of such studies will often quell any vocal (alpha) males that I occasionally come across when giving public lectures. As you may imagine, their ardent, antisleep stance becomes a little wobbly upon receiving such information. With a genuine lack of malice, I proceed to inform them that men who report sleeping too little—or having poor-quality sleep—have a 29 percent lower sperm count than those obtaining a full and restful night of sleep, and the sperm themselves have more deformities. I usually conclude my response with a parenthetical low blow, noting that short sleep duration is significantly associated with smaller-sized testicles.

Low testosterone is a clinically concerning and life-impacting matter. Males with low testosterone often feel tired and fatigued throughout the day. They find it difficult to concentrate on work tasks, as testosterone has a sharpening effect on the brain's ability to focus. And of course, they have a dulled libido, making an active, fulfilling, and healthy sex life more challenging. Indeed, the self-reported mood and vigor of the young men described in the above study progressively decreased in lockstep with their increasing state of sleep deprivation and their declining levels of testosterone. Add to this the fact that testosterone maintains bone density, and plays a causal role in building muscle mass and therefore strength, and you can begin to get a sense of why a full night of sleep—and the natural hormonal replacement therapy it provides—is so essential to this aspect of health and an active life for men of all ages.

Men are not the only ones who become reproductively compromised by a lack of sleep. Routinely sleeping less than six hours a night results in a 20 percent drop in follicular-releasing hormone in women—a critical female reproductive element that peaks just prior to ovulation and is necessary for conception. In a report that brought together findings from studies over the past forty years of more than 100,000 employed women, those working irregular nighttime hours resulting in poor-quality sleep, such as nurses who performed shift work (a profession occupied almost exclusively by women at the time of these earlier studies), had a significantly higher rate of abnormal menstrual cycles than those working regular daytime hours. Moreover, the women working erratic hours were 80 percent more likely to suffer from issues of sub-fertility that reduced the ability to get pregnant. Women who do become pregnant and routinely sleep less than eight hours a night are also more likely to suffer a miscarriage in the first trimester, relative to those consistently sleeping eight hours or more a night.

Combine these deleterious effects on reproductive health in a couple where both parties are lacking in sleep, and it's easy to appreciate why the epidemic of sleep deprivation is linked to infertility or sub-fertility, and why Darwin would find these results so meaningful in the context of future evolutionary success.

Incidentally, should you ask Dr. Tina Sundelin, my friend and colleague at Stockholm University, how attractive you look when sleep-deprived—a physical expression of underlying biology that alters your chances of pair bonding and thus reproduction—she will inform you of an ugly truth. Sundelin isn't the one doing the judging in this scientific beauty contest. Rather, she conducted an elegant experiment in which members of the public did that for her.

Sundelin took a group of healthy men and women ranging from eighteen to thirty-one years old. They were all photographed twice under identical indoor lighting conditions, same time of day (2:30 p.m.), hair down, no makeup for the women, clean-shaven for the men. What differed, however, was the amount of sleep these individuals were allowed to get before each of the photo shoots. In one of the sessions, the participants were given just five hours of sleep before being put in front of the camera, while in the other session, these same individuals got a full eight hours of sleep. The order of these two conditions was randomized as either first or second across the unwitting models.

She brought another group of participants into the laboratory to act as independent judges. These individuals were naïve to the true purpose of the experiment, knowing nothing about the two different sleep manipulations that had been imposed on the people featured in the photographs. The judges viewed both sets of the pictures in a jumbled order and were asked to give ratings on three features: perceived health, tiredness, and attractiveness.

Despite knowing nothing about the underlying premise of the study, thus operating blind to the different sleep conditions, the judges' scores were unambiguous. The faces pictured after one night of short sleep were rated as looking more fatigued, less healthy, and significantly less attractive, compared with the appealing image of that same individual after they had slept a full eight hours. Sundelin had revealed the true face of sleep loss, and with it, ratified the long-held concept of "beauty sleep."

What we can learn from this still burgeoning area of research is that key aspects of the human reproductive system are affected by sleep in both men and women. Reproductive hormones, reproductive organs, and the very nature of physical attractiveness that has a say in reproductive opportunities: all are degraded by short sleeping. One can only imagine Narcissus being a solid eight- to nine-hour sleeper on the basis of the latter association, perhaps with an afternoon nap for good measure, taken beside the reflection pool.

SLEEP LOSS AND THE IMMUNE SYSTEM

Recall the last time you had the flu. Miserable, wasn't it? Runny nose, aching bones, sore throat, heavy cough, and a total lack of energy. You probably just wanted to curl up in bed and sleep. As well you should. Your body is trying to sleep itself well. An intimate and bidirectional association exists between your sleep and your immune system.

Sleep fights against infection and sickness by deploying all manner of weaponry within your immune arsenal, cladding you with protection. When you do fall ill, the immune system actively stimulates the sleep system, demanding more bed rest to help reinforce the war effort.

Reduce sleep even for a single night, and that invisible suit of immune resilience is rudely stripped from your body.

Short of inserting rectal probes to measure core body temperature in certain sleep research studies, my good colleague Dr. Aric Prather at the University of California, San Francisco, has performed one of the most fetid sleep experiments that I am aware of. He measured the sleep of more than 150 healthy men and women for a week using a wristwatch device. Then he quarantined them, and proceeded to squirt a good dose of rhinovirus, or a live culture of the common cold virus, straight up their noses. I should note that all participants knew about this ahead of time, and had surprisingly given full consent to this snout abuse.

Once the flu virus had been satisfactorily boosted up the nostrils of the participants, Prather then kept them in the laboratory for the following week, monitoring them intensely. He not only assessed the extent of immune reaction by taking frequent samples of blood and saliva, but he also gathered nearly every glob of nasal mucus that the participants produced. Prather had the participants regimentally blowing their noses, and every drop of the product was bagged, tagged, weighed, and analytically pored over by his research team. Using these measures—blood and saliva immune antibodies, together with the average amount of snot evacuated by the participants—Prather could determine whether someone had objectively caught a cold.

Prather retrospectively separated the participants into four subgroups on the basis of how much sleep they had obtained in the week before being exposed to the common cold virus: less than five hours of sleep, five to six hours of sleep, six to seven hours of sleep, and seven or more hours of sleep. There was a clear, linear relationship with infection rate. The less sleep an individual was getting in the week before facing the active common cold virus, the more likely it was that they would be infected and catch a cold. In those sleeping five hours on average, the infection rate was almost 50 percent. In those sleeping seven hours or more a night in the week prior, the infection rate was just 18 percent.

Considering that infectious illnesses, such as the common cold, influenza, and pneumonia, are among the leading causes of death in developed countries, doctors and governments would do well to stress the critical importance of sufficient sleep during the flu season.

Perhaps you are one of the responsible individuals who will get a flu shot each year, boosting your own resilience while adding strength to the immunity of the herd—your community. However, that flu shot is only effective if your body actually reacts to it by generating antibodies.

A remarkable discovery in 2002 demonstrated that sleep impacts your response to a standard flu vaccine. In the study, healthy young adults were separated into two groups: one had their sleep restricted to four hours a night for six nights, and the other group was allowed seven and a half to eight and a half hours of time in bed each night. At the end of the six days, everyone was given a flu shot. In the days afterward, researchers took blood samples to determine how effective these individuals were in generating an antibody response, determining whether or not the vaccination was a success.

Those participants who obtained seven to nine hours' sleep in the week before getting the flu shot generated a powerful antibody reaction, reflecting a robust, healthy immune system. In contrast, those in the sleep-restricted group mustered a paltry response, producing less than 50 percent of the immune reaction their well-slept counterparts were able to mobilize. Similar consequences of too little sleep have since been reported for the hepatitis A and B vaccines.

Perhaps the sleep-deprived individuals could still go on to produce a more robust immune reaction if only they were given enough recovery sleep time? It's a nice idea, but a potentially false one. Even if an individual is allowed two or even three weeks of recovery sleep to get over the assault of one week of short sleeping, they never go on to develop a full immune reaction to the flu shot. In fact, a diminution in certain immune cells could still be observed a year later in the participants after just a minor, short dose of sleep restriction. As with the effects of sleep deprivation on memory, once you miss out on the benefit of sleep in the moment—here, regarding an immune response to this season's flu—you cannot regain the benefit simply by trying to catch up on lost sleep. The damage is done, and some of that harm can still be measured a year later.

No matter what immunological circumstance you find yourself in—be it preparation for receiving a vaccine to help boost immunity, or mobilizing a mighty adaptive immune response to defeat a viral attack—sleep, and a full night of it, is especially prudent.

It doesn't require many nights of short sleeping before the body is rendered immunologically weak, and here the issue of cancer becomes relevant. Natural killer cells are an elite and powerful squadron within the ranks of your immune system. Think of natural killer cells like the secret service agents of your body, whose job it is to identify dangerous foreign elements and eliminate them—007 types, if you will.

One such foreign entity that natural killer cells will target are malignant (cancerous) tumor cells. Natural killer cells will effectively punch a hole in the outer surface of these cancerous cells and inject a protein that can destroy the malignancy. What you want, therefore, is a virile set of these James Bond–like immune cells at all times. That is precisely what you don't have when sleeping too little.

Dr. Michael Irwin at the University of California, Los Angeles, has performed landmark studies revealing just how quickly and comprehensively a brief dose of short sleep can affect your cancer-fighting immune cells. Examining healthy young men in a careful study that controlled for spurious factors, such as physical activity, Irwin demonstrated that a single night of four hours of sleep—such as going to bed at three a.m. and waking up at seven a.m.—swept away 70 percent of the natural killer cells circulating in the immune system, relative to a full eight-hour night of sleep. That is a significant state of immune deficiency to find yourself facing, and it happens quickly, after essentially one "bad night" of sleep.

A number of prominent epidemiological studies have reported that nighttime shift work, and the disruption to circadian rhythms and sleep that it causes, up your odds of developing some forms of cancer.

Stirred by the strength of accumulating evidence, Denmark recently became the first country to pay worker compensation to women who had developed breast cancer after years of night-shift work in government-sponsored jobs, such as nurses and air cabin crew. Other governments—Britain, for example—have so far resisted similar legal claims, refusing payout compensation.

Exactly how and why sleep disruption is linked to certain forms of cancer is also becoming clear. Part of the problem relates back to the agitating influence of the sympathetic nervous system as it is forced into overdrive by a lack of sleep. Ramping up the body's level of sympathetic nervous activity will provoke an unnecessary and sustained inflammation response from the immune system. When faced with a real threat, a brief spike of sympathetic nervous system activity will often trigger a similarly transient response from inflammatory activity—one that is useful in anticipation of potential bodily harm (think of a physical tussle with a wild animal or rival hominid tribe). However, inflammation has a dark side. Left switched on without a natural return to peaceful quiescence, a nonspecific state of chronic inflammation causes manifold health problems, including those relevant to cancer.

Cancers are known to use the inflammation response to their advantage. For example, some cancer cells will lure inflammatory factors into the tumor mass to help initiate the growth of blood vessels that feed it with more nutrients and oxygen. Tumors can also use inflammatory factors to help further damage and mutate the DNA of their cancer cells, increasing the tumor's potency. Inflammatory factors associated with sleep deprivation may also be used to help physically shear some of the tumor from its local moorings, allowing the cancer to up-anchor and spread to other territories of the body. It is a state called metastasis, the medical term for the moment when cancer breaches the original tissue boundaries of origin (here, the injection site) and begins to appear in other regions of the body.

It is these cancer-amplifying and -spreading processes that we now know a lack of sleep will encourage, as recent studies by Dr. David Gozal at the University of Chicago have shown. In his study mice were first injected with malignant cells, and tumor progression was then tracked across a four-week period. Half of the mice were allowed to sleep normally during this time; the other half had their sleep partially disrupted, reducing overall sleep quality.

The sleep-deprived mice suffered a 200 percent increase in the speed and size of cancer growth, relative to the well-rested group. When Gozal performed postmortems of the mice, he discovered that the tumors were more aggressive in the sleep-deficient animals. Their cancer had metastasized, spreading to surrounding organs, tissue, and bone. Modern medicine is increasingly adept in its treatment of cancer when it stays put, but when cancer metastasizes—as was encouraged by the state of sleep disruption—medical intervention often becomes challenging, and death rates escalate.

Gozel has further drawn back the curtains of sleep disruption to reveal the mechanisms responsible for this malignant state of affairs. Gozal has shown that immune cells, called tumor-associated macrophages, are one root cause of the cancerous influence of sleep disruption. He found that sleep deprivation will diminish one form of these macrophages, called M1 cells, that otherwise help combat cancer. Yet sleep disruption conversely boosts levels of an alternative form of macrophages, called M2 cells, which promote cancer growth. This combination helped explain the devastating carcinogenic effects seen in the mice when their sleep was manipulated.

Sleep disruption may therefore increase the risk of cancer development and, if cancer is established, favor its rapid and more rampant growth. The scientific evidence linking disrupted sleep-wake rhythms and cancer is now so damning that the World Health Organization has officially classified nighttime shift work as a "probable carcinogen."

SLEEP LOSS, GENES, AND DNA

Beyond increasing your risk for developing Alzheimer's disease, diabetes, depression, obesity, hypertension, and cardiovascular disease, chronic sleep loss can erode the very essence of biological life itself: your genetic code and the structures that encapsulate it.

Each cell in your body has an inner core, or nucleus. Within that nucleus resides most of your genetic material in the form of deoxyribonucleic acid (DNA) molecules. DNA molecules form beautiful helical strands, like tall spiral staircases in an opulent home. Segments of these spirals provide specific engineering blueprints that instruct your cells to perform particular functions. These distinct segments are called genes. Rather like double-clicking open a Word file on your computer and then sending it to your printer, when genes are activated and read by the cell, a biological product is printed out, such as the creation of an enzyme that helps with digestion, or a protein that helps strengthen a memory circuit within the brain.

Anything that causes a shimmy or wobble in gene stability can have consequences. Erroneously over- or under-expressing particular genes can cause biologically printed products that raise your risk of disease, such as dementia, cancer, cardiovascular ill health, and immune dysfunction. Enter the destabilizing force of insufficient sleep.

Thousands of genes within the brain depend upon consistent and sufficient sleep for their stable regulation. Deprive a mouse of sleep for just a day, as researchers have done, and the activity of hundreds of these genes becomes significantly impaired. Like a stubborn file that refuses to be transcribed by a printer, when you do not lavish these DNA segments with enough sleep, they will not translate their instructional code into printed action and give the brain and body what they need.

Dr. Derk-Jan Dijk, who directs the Surrey Sleep Research Center in England, has shown that the effects of insufficient sleep on genetic activity are just as striking in humans as they are in mice. Dijk and his prolific team examined gene expression in a group of healthy young men and women after having restricted them to a six-hour window of sleep a night for one week, all monitored under strict laboratory conditions. After one week of subtly reduced sleep, the activity of a hefty 711 genes was distorted, relative to the genetic activity profile of these very same individuals when they were obtaining an eight-and-a-half-hour sleep opportunity for a week.

Interestingly, the effect went in both directions: about half of those 711 genes had been abnormally revved up in their expression by the loss of sleep, while the other half had been diminished in their expression, or shut down entirely. The genes that were increased included those linked to chronic inflammation, cellular stress, and various factors that cause cardiovascular disease. Among those turned down were genes that help maintain stable metabolism and optimal immune responses. Subsequent studies have found that short sleep duration will also disrupt the activity of genes regulating cholesterol. In particular, a lack of sleep will cause a drop in high-density lipoproteins (HDLs)—a directional profile that has consistently been linked to cardiovascular disease.*

*Beyond a simple lack of sleep, Dijk's research team has further shown that inappropriately timed sleep, such as that imposed by jet lag or shift work, can have equally large effects on the expression of human genes as inadequate sleep. By pushing forward an individual's sleep-wake cycle by a few hours each day for three days, Dijk disrupted a massive

Insufficient sleep does more than alter the activity and readout of your genes; it attacks the physical structure of your genetic material itself. The spiral strands of DNA in your cells float around in the nucleus, but are tightly wound together into structures called chromosomes, rather like weaving individual threads together to make a sturdy shoelace. And just like a shoelace, the ends of your chromosomes need to be protected by a cap or binding tip. For chromosomes, that protective cap is called a telomere. If the telomeres at the end of your chromosomes become damaged, your DNA spirals become exposed and your now vulnerable genetic code cannot operate properly, like a fraying shoelace without a tip.

Short sleep and poor-quality sleep are both associated with greater damage to the capstone telomeres of an individual. These are the findings of a collection of studies that have recently been reported in adults in their forties, fifties, and sixties by numerous independent research teams around the world.*

Whether this association is causal remains to be determined. But the particular nature of the telomere damage caused by short sleeping is now becoming clear. It appears to mimic that seen in aging or advanced decrepitude. That is, two individuals of the same chronological age would not appear to be of the same biological age on the basis of their telomere health if one was routinely sleeping five hours a night while the other was sleeping seven hours a night. The latter would appear "younger," while the former would artificially have aged far beyond their calendar years.

Genetic engineering of animals and genetically modified food are fraught topics, layered with strong emotions. DNA occupies a transcendent, near-divine position in the minds of many individuals, liberal and conservative alike. On this basis, we should feel just as averse and uncomfortable about our own lack of sleep. Not sleeping enough

---

one-third of the transcribing activity of the genes in a group of young, healthy adults. Once again, the genes that were impacted controlled elemental life processes, such as the timing of metabolic, thermoregulatory, and immune activity, as well as cardiac health.

*The significant relationship between short sleep and short or damaged telomeres is observed even when accounting for other factors that are known to harm telomeres, such as age, weight, depression, and smoking.

may modify your gene transcriptome—that is, the very essence of you, or at least you as defined biologically by your DNA. Neglecting sleep could therefore represent a strange form of genetic engineering, tampering with the nucleic alphabet that spells out your daily health story. Permit the same in our children and teenagers, and we may be imposing a similar genetic engineering experiment on them as well.

PART 3

How and Why We Dream

CHAPTER 9

Routinely Psychotic

REM-Sleep Dreaming

Last night, you became flagrantly psychotic. It will happen again tonight. Before you reject this diagnosis, allow me to offer five justifying reasons. First, when you were dreaming last night, you started to see things that were not there—you were *hallucinating*. Second, you believed things that could not possibly be true—you were *delusional*. Third, you became confused about time, place, and person—you were *disoriented*. Fourth, you had extreme swings in your emotions—something psychiatrists call being *affectively labile*. Fifth (and how delightful!), you woke up this morning and forgot most, if not all, of this bizarre dream experience—you were suffering from *amnesia*. If you were to experience any of these symptoms while awake, you'd be seeking immediate psychological treatment. Yet for reasons that are only now becoming clear, the brain state called REM sleep and the mental experience that goes along with it, dreaming, are normal biological and psychological processes, and truly essential ones, as we shall learn.

REM sleep is not the only time during sleep when we dream. Indeed, if you use a liberal definition of dreaming as any mental activity reported upon awakening from sleep, such as "I was thinking about rain," then you technically dream in all stages of sleep. If I wake you from the deepest stage of NREM sleep, there is a 0 to 20 percent chance you will report some type of bland thought like this. As you are falling asleep or exiting sleep, the dream-like experiences you have tend to be visually or movement based. But dreams as most of us think of them—those hallucinogenic, motoric, emotional, and bizarre experiences with a rich narrative—come from REM sleep, and many sleep researchers limit their definition of true dreaming to that which occurs in REM sleep. As a result, this chapter will mainly focus on REM sleep and the dreams that emerge from this state. We will, however, still explore dreaming at these other moments of sleep, as those dreams, too, offer important insights into the process itself.

YOUR BRAIN ON DREAMS

In the 1950s and 1960s, recordings using electrodes placed on the scalp gave scientists a general sense of the type of brainwave activity underpinning REM sleep. But we had to wait until the advent of brain-imaging machines in the early 2000s before we could reconstruct glorious, three-dimensional visualizations of brain activity during REM sleep. It was worth the wait.

Among other breakthroughs, the method and the results undermined the postulates of Sigmund Freud and his nonscientific theory of dreams as wish fulfillment, which had dominated psychiatry and psychology for an entire century. There were important virtues of Freud's theory, and we will discuss them below. But there were deep and systemic flaws that led to a rejection of the theory by modern-day science. Our more informed, neuroscientific view of REM sleep has since given rise to scientifically testable theories of *how* it is that we dream (e.g., logical/illogical, visual/non-visual, emotional/non-emotional) and *what* it is that we dream about (e.g., experiences from our recent waking lives/de novo experiences), and even gives the chance to nibble away at surely the most fascinating question in all of sleep science—and arguably science writ large—*why* it is that we dream, that is, the function(s) of REM-sleep dreaming.

To appreciate the advance that brain scanners made to our understanding of REM sleep and dreaming beyond simple EEG recordings, we can return to our sports stadium analogy from chapter 3. Dangling a microphone over the stadium can measure the summed activity of the entire crowd. But it is geographically nonspecific in this regard. You cannot determine whether one segment of the crowd in the stadium is chanting loudly while the segment directly next door is relatively less vocal, or even completely silent.

The same nonspecificity is true when measuring brain activity with an electrode placed on the scalp. However, magnetic resonance imaging (MRI) scans do not suffer this same spatial smearing effect in quantifying brain activity. MRI scanners effectively carve up the stadium (the brain) into thousands of small, discrete boxes, rather like individual pixels on a screen, and then measure the local activity of the crowd (brain cells) within that specific pixel, distinct from other pixels in other parts of the stadium. Furthermore, MRI scanners map this activity in three dimensions, covering all levels of the stadium brain—lower, middle, upper.

By placing individuals inside brain scanning machines, I and many other scientists have been able to observe the startling changes in brain activity that occur when people enter into REM sleep and begin dreaming. For the first time, we could see how even the very deepest structures previously hidden from view came alive as REM sleep and dreaming got under way.

During dreamless, deep NREM sleep, overall metabolic activity shows a modest decrease relative to that measured from an individual while they are resting but awake. However, something very different happens as the individual transitions into REM sleep and begins to dream. Numerous parts of the brain "light up" on the MRI scan as REM sleep takes hold, indicating a sharp increase in underlying activity. In fact, there are four main clusters of the brain that spike in activity when someone starts dreaming in REM sleep: (1) the visuospatial regions at the back of the brain, which enable complex visual perception; (2) the motor cortex, which instigates movement; (3) the hippocampus and surrounding regions that we have spoken about before, which support your autobiographical memory; and (4) the deep emotional centers of the brain—the amygdala and the cingulate cortex, a ribbon of tissue that sits above the amygdala and lines the inner surface of your brain—both of which help generate and process emotions. Indeed, these emotional regions of the brain are up to 30 percent more active in REM sleep compared to when we are awake!

Since REM sleep is associated with the active, conscious experience of dreaming, it was perhaps predictable that REM sleep would involve a similarly enthusiastic pattern of increased brain activity. What came as a surprise, however, was a pronounced *deactivation* of other brain regions—specifically, circumscribed regions of the far left and right sides of the prefrontal cortex. To find this area, take your hands and place them at the side corners of the front of your head, about two inches above the corners of your eyes (think of the crowd's universal hand placement when a player just misses scoring a goal during overtime in a World Cup soccer game). These are the regions that became icy blue color scheme blobs on the brain scans, informing us that these neural territories had become markedly suppressed in activity during the otherwise highly active state of REM sleep.

Discussed in chapter 7, the prefrontal cortex acts like the CEO of the brain. This region, especially the left and right sides, manages rational thought and logical decision-making, sending "top-down" instructions to your more primitive deep-brain centers, such as those instigating emotions. And it is this CEO region of your brain, which otherwise maintains your cognitive capacity for ordered, logical thought, that is temporarily ousted each time you enter into the dreaming state of REM sleep.

REM sleep can therefore be considered as a state characterized by strong activation in visual, motor, emotional, and autobiographical memory regions of the brain, yet a relative deactivation in regions that control rational thought. Finally, thanks to MRI, we had our first scientifically grounded, whole-brain visualization of the brain in REM sleep. Coarse and rudimentary as the method was, we entered a new era of understanding the *why* and the *how* of REM-sleep dreaming, without relying on idiosyncratic rules or opaque explanations of past dream theories, such as Freud's.

We could make simple, scientific predictions that could be falsified or supported. For example, after having measured the pattern of brain activity of an individual in REM sleep, we could wake them up and obtain a dream report. But even without that dream report, we should be able to read the brain scans and accurately predict the nature of that person's dream before they report it to us. If there was minimal motor activity, but a lot of visual and emotional brain activity, then the particular dream should have little movement but be filled with visual objects and scenes and contain strong emotions—and vice versa. We have conducted just such an experiment, and the findings were so: we could predict with confidence the *form* of someone's dream—would it be visual, would it be motoric, would it be awash with emotion, would it be completely irrational and bizarre?—before the dreamers themselves reported their dream experience to the research assistant.

As revolutionary as it was to predict the general *form* of someone's dream (emotional, visual, motoric, etc.), it left a more fundamental question unanswered: Can we predict the *content* of someone's dream—that is, can we predict *what* an individual is dreaming about (e.g., a car, a woman, food), rather than just the *nature* of the dream (e.g., is it visual)?

In 2013, a research team in Japan, led by Dr. Yukiyasu Kamitani at the Advanced Telecommunications Research Institute International in Kyoto, found an ingenious way to address the question. They essentially cracked the code of an individual's dream for the very first time and, in doing so, led us to an ethically uncomfortable place.

Individuals in the experiment consented to the study—an important fact, as we shall see. The results remain preliminary, since they were obtained in just three individuals. But they were highly significant. Also, the researchers focused on the short dreams we all frequently have just at the moment when we are falling asleep, rather than the dreams of REM sleep, though the method will soon be applied to REM sleep.

The scientists placed each participant into an MRI scanner numerous times over the course of several days. Every time the participant fell asleep, the researchers would wait for a short while as they recorded the brain activity, and then wake the person up and obtain a dream report. Then they would let the person fall back to sleep, and repeat the procedure. The researchers continued to do this until they had gathered hundreds of dream reports and corresponding snapshots of brain activity from their participants. An example of one of the dream reports was: "I saw a big bronze statue . . . on a small hill, and below the hill there were houses, streets, and trees."

Kamitani and his team then distilled all of the dream reports down into twenty core content categories that were most frequent in the dreams of these individuals, such as books, cars, furniture, computers, men, women, and food. To obtain some kind of ground truth of what participants' brain activity looked like when they actually perceived these types of visual images while awake, the researchers selected real photographs that represented each category (relevant pictures of cars, men, women, furniture, etc.). Participants were then placed back inside the MRI scanner and shown these images while awake as the researchers measured their brain activity again. Then, using these patterns of waking brain activity as a truth template of sorts, Kamitani went pattern-matching in the sea of sleeping brain activity. The concept is somewhat like DNA matching at a crime scene: the forensics team obtains a sample of the victim's DNA that they use as a template, then go in search of a specific match from among the myriad possible samples.

The scientists were able to predict with significant accuracy the content of participants' dreams at any one moment in time using just the MRI scans, operating completely blind to the dream reports of the participants. Using the template data from the MRI images, they could tell if you were dreaming of a man or a woman, a dog or a bed, flowers or a knife. They were, in effect, mind reading, or should I say, dream reading. The scientists had turned the MRI machine into a very expensive version of the beautiful handmade dream-catchers that some Native American cultures will hang above their beds in the hopes of ensnaring the dream—and they had succeeded.

The method is far from perfect. It cannot currently determine exactly what man, woman, or car the dreamer is seeing. For example, a recent dream of my own shamelessly featured a stunning 1960s vintage Aston Martin DB4, though you'd never be able to determine that degree of specificity from MRI scans, should I have been a participant in the experiment. You would simply know that I was dreaming of *a* car rather than, say, a computer or piece of furniture, but not *which* car it was. Nevertheless, it is a remarkable advance that will only improve to the point of scientists having the clear ability to decode and visualize dreams. We can now begin to learn more about the construction of dreams, and that knowledge may help disorders of the mind in which dreams are deeply problematic, such as trauma nightmares in PTSD patients.

As an individual, rather than a scientist, I must admit to having some vague unease with the idea. Once, our dreams were our own. We got to decide whether or not to share them with others and, if we did, which parts to include and which parts to withhold. Participants in these studies always give their consent. But will the method someday reach beyond science and into the philosophical and ethical realm? There may well be a time in the not-too-distant future where we can accurately "read out" and thus take ownership of a process that few people have volitional control over—the dream.* When this finally happens, and I'm sure it will, do we hold the dreamer responsible for what they dream? Is it fair to judge what it is they are dreaming, since they were not the conscious architect of their dream? But if they were not, then who is? It is a perplexing and uncomfortable issue to face.

THE MEANING AND CONTENT OF DREAMS

MRI studies helped scientists better understand the nature of dreaming, and allowed low-level decoding of dreams. Results of these brain scanning experiments have also led to a prediction about one of the oldest questions in all of humanity, and certainly of sleep: Where do dreams come from?

Before the new science of dreaming, and before Freud's unsystematic treatment of the topic, dreams came from all manner of sources. The ancient Egyptians believed dreams were sent down from the gods on high. The Greeks shared a similar contention, regarding dreams as visitations from the gods, offering information divine. Aristotle, however, was a notable exception in this regard. Three of the seven topics in his *Parva Naturalia* (Short Treatises on Nature) addressed the state of slumber: *De Somno et Vigilia* (On Sleep), *De Insomniis* (On Dreams),

*I say few, since there are some individuals who can not only become aware that they are dreaming, but even control how and what they dream. It is called lucid dreaming, and we shall read much more about it in a later chapter.

and *De Divinatione per Somnum* (On Divination in Sleep). Levelheaded as always, Aristotle dismissed the idea of dreams as being heavenly directed, and instead he cleaved strongly to the more self-experienced belief that dreams have their origins in recent waking events.

But it was actually Freud who, in my opinion, made the most remarkable scientific contribution to the field of dream research, one that I feel modern-day neuroscience does not give him sufficient credit for. In his seminal book *The Interpretation of Dreams* (1899), Freud situated the dream unquestionably within the brain (that is, the mind, as there is arguably no ontological difference between the two) of an individual. That may seem obvious now, even inconsequential, but at the time it was anything but, especially considering the aforementioned past. Freud had single-handedly wrested dreams from the ownership of celestial beings, and from the anatomically unclear location of the soul. In doing so, Freud made dreams a clear domain of what would become neuroscience—that is, the terra firma of the brain. True and inspired was his proposal that dreams emerge from the brain, as it implied that answers could only be found by way of a systematic interrogation of the brain. We must thank Freud for this paradigmatic shift in thinking.

Yet Freud was 50 percent right and 100 percent wrong. Things quickly went downhill from this point, as the theory plunged into a quagmire of unprovability. Simply put, Freud believed that dreams came from unconscious wishes that had not been fulfilled. According to his theory, repressed desires, which he termed the "latent content," were so powerful and shocking that if they appeared in the dream undisguised, they would wake the dreamer up. To protect the dreamer and his sleep, Freud believed there was a censor, or a filter, within the mind. Repressed wishes would pass through the censor and emerge disguised on the other side. The camouflaged wishes and desires, which Freud described as the "manifest content," would therefore be unrecognizable to the dreamer, carrying no risk of jolting the sleeping individual awake.

Freud believed that he understood how the censor worked and that, as a result, he could decrypt the disguised dream (manifest content) and reverse-engineer it to reveal the true meaning (latent content, rather like email encryption wherein the message is cloaked with a code). Without the decryption key, the content of the email cannot be read. Freud felt that he had discovered the decryption key to everyone's dreams, and for many of his affluent Viennese patients, he offered the paid service of removing this disguise and revealing to them the original message content of their dreams.

The problem, however, was the lack of any clear predictions from Freud's theory. Scientists could not design an experiment that would test any tenets of his theory in order to help support or falsify it. It was Freud's genius, and his simultaneous downfall. Science could never prove him wrong, which is why Freud continues to cast a long shadow on dream research to this day. But by the very same token, we could never prove the theory right. A theory that cannot be discerned true or false in this way will always be abandoned by science, and that is precisely what happened to Freud and his psychoanalytic practices.

As a concrete example, consider the scientific method of carbon dating, used to determine the age of an organic object like a fossil. To validate the method, scientists would have the same fossil analyzed by several different carbon-dating machines that operated on the same underlying principle. If the method was scientifically robust, these independent machines should all return the same value of the fossil's age. If they do not, the method must be flawed, as the data is inaccurate and cannot be replicated.

The method of carbon dating was shown by this process to be legitimate. Not so for the Freudian psychoanalytic method of dream interpretation. Researchers have had different Freudian psychoanalysts interpret the same dream of an individual. If the method was scientifically reliable, with clear structured rules and metrics that the therapists could apply, then their respective interpretations of this dream should be the same—or at least have some degree of similarity in the extracted meaning they return. Instead, the psychoanalysts all gave remarkably different interpretations of this same dream, without any statistically significant similarity between them. There was no consistency. You cannot place a "QC"—quality control—sticker on Freudian psychoanalysis.

A cynical criticism of the Freudian psychoanalytic method is therefore one of "the disease of generic-ness." Rather like horoscopes, the interpretations offered are generalizable, seemingly providing an explanatory fit to any and all things. For example, before describing the criticisms of Freudian theory in my university lectures, I often do the following with my students as a (perhaps cruel) demonstration. I start by asking anyone in the lecture auditorium if they would be willing to share a dream that I will interpret pro bono, on the spot. A few hands will go up. I point to one of the respondents and ask them their name—let's call this one Kyle. I ask Kyle to tell me his dream. He says:

> I was running through an underground parking lot trying to find my car. I don't know why I was running, but I felt like I really needed to get to my car. I found the car, um, but it wasn't actually the car I owned but I thought it was my car in the dream. I tried to start the car, but each time I turned the key, nothing happened. Then my cell phone went off loudly and I woke up.

In response, I look intensely and knowingly at Kyle, having been nodding my head throughout his description. I pause, and then say, "I know *exactly* what your dream is about, Kyle." Amazed, he (and the rest of the lecture hall) awaits, my answer as though time has ground to a halt. After another long pause, I confidently enunciate the following: "Your dream, Kyle, is about time, and more specifically, about not having enough time to do the things you really want to do in life." A wave of recognition, almost relief, washes over Kyle's face, and the rest of the class appear equally convinced.

Then I come clean. "Kyle—I have a confession. No matter what dream anyone ever tells me, I always give them that very same generic response, and it always seems to fit." Thankfully, Kyle is a good sport and takes this with no ill grace, laughing with the rest of the class. I apologize once again to him. The exercise, however, importantly reveals the dangers of generic interpretations that feel very personal and uniquely individual, yet scientifically hold no specificity whatsoever.

I want to be clear, as this all seems dismissive. I am in no way suggesting that reviewing your dreams yourself, or sharing them with someone else, is a waste of time. On the contrary, I think it is a very helpful thing to do, as dreams do have a function, as we will read about in the next chapter. Indeed, journaling your waking thoughts, feelings, and concerns has a proven mental health benefit, and the same appears true of your dreams. A meaningful, psychologically healthy life is an examined one, as Socrates so often declared. Nevertheless, the psychoanalytic method built on Freudian theory is nonscientific and holds no repeatable, reliable, or systematic power for decoding dreams. This, people must be made aware of.

In actual fact, Freud knew of this limitation. He had the prophetic sensibility to recognize that a day of scientific reckoning would come. The sentiment is neatly encapsulated in his own words when discussing the origin of dreams in *The Interpretation of Dreams*, where he states: "deeper research will one day trace the path further and discover an organic basis for the mental event." He knew that an organic (brain) explanation would ultimately reveal the truth of dreams—a truth that his theory lacked.

Indeed, four years before he descended into a nonscientific, psychoanalytic theory of dreaming in 1895, Freud initially tried to construct a scientifically informed, neurobiological explanation of the mind in a work called the *Project for a Scientific Psychology*. In it are beautiful drawings of neural circuits with connecting synapses that Freud mapped out, trying to understand the workings of the mind while awake and asleep. Unfortunately, the field of neuroscience was still in its infancy at the time. Science was simply not up to the task of deconstructing dreams, and so unscientific postulates such as Freud's were inevitable. We should not blame him for that, but we should also not accept an unscientific explanation of dreams *because* of that.

Brain scanning methods have offered the first inklings of just this organic truth about the source of dreams. Since autobiographical memory regions of the brain, including the hippocampus, are so active during REM sleep, we should expect dreaming to contain elements of the individual's recent experience and perhaps give clues as to the meaning, if any, of dreams: something that Freud elegantly described as "day residue." It was a clear-cut, testable prediction, which my longtime friend and colleague Robert Stickgold at Harvard University elegantly proved was, in fact, utterly untrue . . . with an important caveat.

Stickgold designed an experiment that would determine the extent to which dreams were a precise replay of our recent waking autobiographical experiences. For two weeks straight, he had twenty-nine healthy young adults keep a detailed log of daytime activities, the events they were engaged in (going to work, meeting specific friends, meals they ate, sports they played, etc.), and their current emotional concerns. In addition, he had them keep dream journals, asking them to write down any recalled dreams that they had when they woke up each morning. He then had external judges systematically compare the reports of the participants' waking activities with their dream reports, focusing on the degree of similarity of well-defined features, such as locations, actions, objects, characters, themes, and emotions.

Of a total of 299 dream reports that Stickgold collected from these individuals across the fourteen days, a clear rerun of prior waking life events—day residue—was found in just 1 to 2 percent. Dreams are not, therefore, a wholesale replay of our waking lives. We do not simply rewind the video of the day's recorded experience and relive it at night, projected on the big screen of our cortex. If there is such a thing as "day residue," there are but a few drops of the stuff in our otherwise arid dreams.

But Stickgold did find a strong and predictive daytime signal in the static of nighttime dream reports: emotions. Between 35 and 55 percent of emotional themes and concerns that participants were having while they were awake during the day powerfully and unambiguously resurfaced in the dreams they were having at night. The commonalities were just as clear to the participants themselves, who gave similarly confident judgments when asked to compare their own dream reports with their waking reports.

If there is a red-thread narrative that runs from our waking lives into our dreaming lives, it is that of emotional concerns. Counter to Freudian assumptions, Stickgold had shown that there is no censor, no veil, no disguise. Dream sources are transparent—clear enough for anyone to identify and recognize without the need for an interpreter.

DO DREAMS HAVE A FUNCTION?

Through a combination of brain activity measures and rigorous experimental testing, we have finally begun to develop a scientific understanding of human dreams: their form, content, and the waking source(s). There is, however, something missing here. None of the studies that I have described so far proves that dreams have any function. REM sleep, from which principal dreams emerge, certainly has many functions, as we have discussed and will continue to discuss. But do dreams themselves, above and beyond REM sleep, actually do anything for us? As a matter of scientific fact, yes, they do.

CHAPTER 10

Dreaming as Overnight Therapy

It was long thought that dreams were simply epiphenomena of the stage of sleep (REM) from which they emerge. To illustrate the concept of epiphenomena, let's consider the lightbulb.

The reason we construct the physical elements of a lightbulb—the glass sphere, the coiled wire element that sits inside, the screw-in electrical contact at the base—is to create light. That is the function of the lightbulb, and the reason we designed the apparatus to begin with. However, a lightbulb also produces heat. Heat is not the function of the lightbulb, nor is it the reason we originally fashioned it. Instead, heat is simply what happens when light is generated in this way. It is an unintended by-product of the operation, not the true function. Heat is an epiphenomenon in this case.

Similarly, evolution may have gone to great lengths to construct the neural circuits in the brain that produce REM sleep and the functions that REM sleep supports. However, when the (human) brain produces REM sleep in this specific way, it may also produce this thing we call dreaming. Dreams, like heat from a lightbulb, may serve no function. Dreams may simply be epiphenomena of no use or consequence. They are merely an unintended by-product of REM sleep.

Rather a depressing thought, isn't it? I'm sure many of us feel that our dreams have meaning and some useful purpose.

To address this stalemate, exploring whether dreaming, beyond the stage of sleep it emerges from, has true purpose, scientists began by defining the functions of REM sleep. Once those functions were known, we could then examine whether the dreams that accompany REM sleep—and the very specific content of those dreams—were crucial determinants of those adaptive benefits. If what you dream about offers no predictive power in determining the benefits of that REM sleep, it would suggest that dreams are epiphenomenal, and REM sleep alone is sufficient. If, however, you need both REM sleep *and* to be dreaming about specific things to accomplish such functions, it would suggest that REM sleep alone, although necessary, is not sufficient. Rather, a unique combination of REM sleep *plus* dreaming, and dreaming of very particular experiences, is needed to transact these nighttime benefits. If this was proven, dreams could not be dismissed as an epiphenomenal by-product of REM sleep. Rather, science would have to recognize dreaming as an essential part of sleep and the adaptive advantages it supports, above and beyond REM sleep itself.

Using this framework, we have found two core benefits of REM sleep. Both functional benefits require not just that you have REM sleep, but that you dream, and dream about specific things. REM sleep is necessary, but REM sleep alone is not sufficient. Dreams are not the heat of the lightbulb—they are no by-product.

The first function involves nursing our emotional and mental health, and is the focus of this chapter. The second is problem solving and creativity, the power of which some individuals try to harness more fully by controlling their dreams, which we treat in the next chapter.

DREAMING—THE SOOTHING BALM

It is said that time heals all wounds. Several years ago I decided to scientifically test this age-old wisdom, as I wondered whether an amendment was in order. Perhaps it was not time that heals all wounds, but rather time spent in dream sleep. I had been developing a theory based on the combined patterns of brain activity and brain neurochemistry of REM sleep, and from this theory came a specific prediction: REM-sleep dreaming offers a form of overnight therapy. That is, REM-sleep dreaming takes the painful sting out of difficult, even traumatic, emotional episodes you have experienced during the day, offering emotional resolution when you awake the next morning.

At the heart of the theory was an astonishing change in the chemical cocktail of your brain that takes place during REM sleep. Concentrations of a key stress-related chemical called noradrenaline are completely shut off within your brain when you enter this dreaming sleep state. In fact, REM sleep is the only time during the twenty-four-hour period when your brain is completely devoid of this anxiety-triggering molecule. Noradrenaline, also known as norepinephrine, is the brain equivalent to a body chemical you already know and have felt the effects of: adrenaline (epinephrine).

Previous MRI studies established that key emotion- and memory-related structures of the brain are all reactivated during REM sleep, as we dream: the amygdala and emotion-related regions of the cortex, and the key mnemonic center, the hippocampus. Not only did this suggest the possibility that emotion-specific memory processing was possible, if not probable, during the dreaming state, but now we understood that this emotional memory reactivation was occurring in a brain free of a key stress chemical. I therefore wondered whether the brain during REM sleep was reprocessing upsetting memory experiences and themes in this neurochemically calm (low noradrenaline), "safe" dreaming brain environment. Is the REM-sleep dreaming state a perfectly designed nocturnal soothing balm—one that removes the emotional sharp edges of our daily lives? It seemed so from everything neurobiology and neurophysiology was telling us (me). If so, we should awake feeling better about distressing events of the day(s) prior.

This was the theory of overnight therapy. It postulated that the process of REM-sleep dreaming accomplishes two critical goals: (1) sleeping to *remember* the details of those valuable, salient experiences, integrating them with existing knowledge and putting them into autobiographical perspective, yet (2) sleeping to *forget*, or dissolve, the visceral, painful emotional charge that had previously been wrapped around those memories. If true, it would suggest that the dream state supports a form of introspective life review, to therapeutic ends.

Think back to your childhood and try to recall some of the strongest memories you have. What you will notice is that almost all of them will be memories of an emotional nature: perhaps a particularly frightening experience of being separated from your parents, or almost being hit by a car on the street. Also notice, however, that your recall of these detailed memories is no longer accompanied by the same degree of emotion that was present at the time of the experience. You have not forgotten the memory, but you have cast off the emotional charge, or at least a significant amount of it. You can accurately relive the memory, but you do not regurgitate the same visceral reaction that was present and imprinted at the time of the episode.* The theory argued that we have REM-sleep dreaming to thank for this palliative dissolving of emotion from experience. Through its therapeutic work at night, REM sleep performed the elegant trick of divorcing the bitter emotional rind from the information-rich fruit. We can therefore learn and usefully recall salient life events without being crippled by the emotional baggage that those painful experiences originally carried.

Indeed, I argued that if REM sleep did not perform this operation, we'd all be left with a state of chronic anxiety in our autobiographical memory networks; every time we recalled something salient, not only would we recall the details of the memory, but we would relive the same stressful emotional charge all over again. Based on its unique brain activity and neurochemical composition, the dream stage of REM sleep helps us avoid this circumstance.

That was the theory, those were the predictions; next came the experimental test, the results of which would take a first step toward falsifying or supporting both.

We recruited a collection of healthy young adults and randomly assigned them to two groups. Each group viewed a set of emotional images while inside an MRI scanner as we measured their emotional brain reactivity. Then, twelve hours later, the participants were placed back inside the MRI scanner and we again presented those same emotional images, cuing their recollection while again measuring emotional brain reactivity. During these two exposure sessions, separated by twelve hours, participants also rated how emotional they felt in response to each image.

Importantly, however, half of the participants viewed the images in the morning and again in the evening, being awake between the two

*An exception is the condition of post-traumatic stress disorder (PTSD), which we will discuss later in this chapter.

viewings. The other half of the participants viewed the images in the evening and again the next morning after a full night of sleep. In this way, we could measure what their brains were objectively telling us using the MRI scans, and in addition, what participants themselves were subjectively feeling about the relived experiences, having had a night of sleep in between, or not.

Those who slept in between the two sessions reported a significant decrease in how emotional they were feeling in response to seeing those images again. In addition, results of the MRI scans showed a large and significant reduction in reactivity in the amygdala, that emotional center of the brain that creates painful feelings. Moreover, there was a reengagement of the rational prefrontal cortex of the brain after sleep that was helping maintain a dampening brake influence on emotional reactions. In contrast, those who remained awake across the day without the chance to sleep and digest those experiences showed no such dissolving of emotional reactivity over time. Their deep emotional brain reactions were just as strong and negative, if not more so, at the second viewing compared with the first, and they reported a similarly powerful reexperiencing of painful feelings to boot.

Since we had recorded the sleep of each participant during the intervening night between the two test sessions, we could answer a follow-up question: Is there something about the type or quality of sleep that an individual experiences that predicts how successful sleep is at accomplishing next-day emotional resolution?

As the theory predicted, it was the dreaming state of REM sleep—and specific patterns of electrical activity that reflected the drop in stress-related brain chemistry during the dream state—that determined the success of overnight therapy from one individual to the next. It was not, therefore, time per se that healed all wounds, but instead it was time spent in dream sleep that was providing emotional convalescence. To sleep, perchance to heal.

Sleep, and specifically REM sleep, was clearly needed in order for us to heal emotional wounds. But was the act of dreaming during REM sleep, and even dreaming of those emotional events themselves, necessary to achieve resolution and keep our minds safe from the clutches of anxiety and reactive depression? This was the question that Dr. Rosalind Cartwright at Rush University in Chicago elegantly dismantled in a collection of work with her clinical patients.

Cartwright, who I contend is as much a pioneer in dream research as Sigmund Freud, decided to study the dream content of people who were showing signs of depression as a consequence of incredibly difficult emotional experiences, such as devastating breakups and bitter divorces. Right around the time of the emotional trauma, she started collecting their nightly dream reports and sifted through them, hunting for clear signs of the same emotional themes emerging in their dream lives relative to their waking lives. Cartwright then performed follow-up assessments up to one year later, determining whether the patients' depression and anxiety caused by the emotional trauma were resolved or continued to persist.

In a series of publications that I still revisit with admiration to this day, Cartwright demonstrated that it was only those patients who were expressly dreaming about the painful experiences around the time of the events who went on to gain clinical resolution from their despair, mentally recovering a year later as clinically determined by having no identifiable depression. Those who were dreaming, but not dreaming of the painful experience itself, could not get past the event, still being dragged down by a strong undercurrent of depression that remained.

Cartwright had shown that it was not enough to have REM sleep, or even generic dreaming, when it comes to resolving our emotional past. Her patients required REM sleep with dreaming, but dreaming of a very specific kind: that which expressly involved dreaming about the emotional themes and sentiments of the waking trauma. It was only that content-specific form of dreaming that was able to accomplish clinical remission and offer emotional closure in these patients, allowing them to move forward into a new emotional future, and not be enslaved by a traumatic past.

Cartwright's data offered further psychological affirmation of our biological overnight therapy theory, but it took a chance meeting at a conference one inclement Saturday in Seattle before my own basic research and theory would be translated from bench to bedside, helping to resolve the crippling psychiatric condition of post-traumatic stress disorder (PTSD).

Patients with PTSD, who are so often war veterans, have a difficult time recovering from horrific trauma experiences. They are frequently plagued by daytime flashbacks of these intrusive memories and suffer reoccurring nightmares. I wondered whether the REM-sleep overnight therapy mechanism we had discovered in healthy individuals had broken down in people suffering from PTSD, thereby failing to help them deal with their trauma memories effectively.

When a veteran soldier suffers a flashback triggered by, say, a car backfiring, they can relive the whole visceral traumatic experience again. It suggested to me that the emotion had not been properly stripped away from the traumatic memory during sleep. If you interview PTSD patients in the clinic, they will often tell you that they just cannot "get over" the experience. In part, they are describing a brain that has not detoxed the emotion from the trauma memory, such that every time the memory is relived (the flashback), so, too, is the emotion, which has not been effectively removed.

Already, we knew that the sleep, especially the REM sleep, of patients suffering from PTSD was disrupted. There was also evidence suggesting that PTSD patients had higher-than-normal levels of noradrenaline released by their nervous system. Building on our overnight therapy theory of REM-sleep dreaming and the emerging data that supported it, I wrote a follow-up theory, applying the model to PTSD. The theory proposed that a contributing mechanism underlying the PTSD is the excessively high levels of noradrenaline within the brain that blocks the ability of these patients from entering and maintaining normal REM-sleep dreaming. As a consequence, their brain at night cannot strip away the emotion from the trauma memory, since the stress chemical environment is too high.

Most compelling to me, however, were the repetitive nightmares reported in PTSD patients—a symptom so reliable that it forms part of the list of features required for a diagnosis of the condition. If the brain cannot divorce the emotion from memory across the first night following a trauma experience, the theory suggests that a repeat attempt of emotional memory stripping will occur on the second night, as the strength of the "emotional tag" associated with the memory remains too high. If the process fails a second time, the same attempt will continue to repeat the next night, and the next night, like a broken record. This was precisely what appeared to be happening with the recurring nightmares of the trauma experience in PTSD patients.

A testable prediction emerged: if I could lower the levels of noradrenaline in the brains of PTSD patients during sleep, thereby reinstating the right chemical conditions for sleep to do its trauma therapy work, then I should be able to restore healthier quality REM sleep. With that restored REM-sleep quality should come an improvement in the clinical symptoms of PTSD, and further, a decrease in the frequency of painful repetitive nightmares. It was a scientific theory in search of clinical evidence. Then came the wonderful stroke of serendipity.

Soon after my theoretical paper was published, I met Dr. Murray Raskind, a remarkable physician who worked at a US Department of Veterans Affairs hospital in the Seattle area. We were both presenting our own research findings at a conference in Seattle and, at the time, we were each unaware of the other's emerging new research data. Raskind—a tall man with kindly eyes whose disarmingly relaxed, jocular demeanor belies a clinical acumen that is not to be underestimated—is a prominent research figure in both the PTSD and Alzheimer's disease fields. At the conference, Raskind presented recent findings that were perplexing to him. In his PTSD clinic, Raskind had been treating his war veteran patients with a generic drug called prazosin to manage their high blood pressure. While the drug was somewhat effective for lowering blood pressure in the body, Raskind found it had a far more powerful yet entirely unexpected benefit within the brain: it alleviated the reoccurring nightmares in his PTSD patients. After only a few weeks of treatment, his patients would return to the clinic and, with puzzled amazement, say things like, "Doc, it's the strangest thing, my dreams don't have those flashback nightmares anymore. I feel better, less scared to fall asleep at night."

It turns out that the drug prazosin, which Raskind was prescribing simply to lower blood pressure, also has the fortuitous side effect of suppressing noradrenaline in the brain. Raskind had delightfully and inadvertently conducted the experiment I was trying to conceive of myself. He had created precisely the neurochemical condition—a lowering of the abnormally high concentrations of stress-related noradrenaline—within the brain during REM sleep that had been absent for so long in these PTSD patients. Prazosin was gradually lowering the harmful high tide of noradrenaline within the brain, giving these patients healthier REM-sleep quality. With healthy REM sleep came a reduction in the patients' clinical symptoms and, most critically, a decrease in the frequency of their repetitive nightmares.

Raskind and I continued our communications and scientific discussions throughout that conference. He subsequently visited my lab at UC Berkeley in the months that followed, and we talked nonstop throughout the day and into the evening over dinner about my neurobiological model of overnight emotional therapy, and how it seemed to perfectly explain his clinical findings with prazosin. These were hairs-on-the-back-of-your-neck-standing-up conversations, perhaps the most exciting I have ever experienced in my career. The basic scientific theory was no longer in search of clinical confirmation. The two had found each other one sky-leaking day in Seattle.

Mutually informed by each other's work, and based on the strength of Raskind's studies and now several large-scale independent clinical trials, prazosin has now been used by the VA for the treatment of repetitive trauma nightmares, and has since received approval by the US Food and Drug Administration for the same benefit.

Many questions remain to be addressed, including more independent replication of the findings in other types of trauma, such as sexual abuse or violence. It is also not a perfect medication due to side effects at higher doses, and not every individual responds to the treatment with the same success. But it is a start. We now have a scientifically informed explanation of one function of REM sleep and the dreaming process inherent in it, and from that knowledge we have taken the first steps toward treating the distressing and disabling clinical condition of PTSD. It may also unlock new treatment avenues regarding sleep and other mental illness, including depression.

DREAMING TO DECODE WAKING EXPERIENCES

Just when I thought REM sleep had revealed all it could offer to our mental health, a second emotional brain advantage gifted by REM sleep came to light—one that is arguably more survival-relevant.

Accurately reading expressions and emotions of faces is a prerequisite of being a functional human being, and indeed, a functional higher primate of most kinds. Facial expressions represent one of the most important signals in our environment. They communicate the emotional state and intent of an individual and, if we interpret them correctly, influence our behavior in return. There are regions of your brain whose job it is to read and decode the value and meaning of emotional signals, especially faces. And it is that very same essential set of brain regions, or network, that REM sleep recalibrates at night.

In this different and additional role, we can think of REM sleep like a master piano tuner, one that readjusts the brain's emotional instrumentation at night to pitch-perfect precision, so that when you wake up the next morning, you can discern overt and subtly covert micro-expressions with exactitude. Deprive an individual of their REM-sleep dreaming state, and the emotional tuning curve of the brain loses its razor-sharp precision. Like viewing an image through frosted glass, or looking at an out-of-focus picture, a dream-starved brain cannot accurately decode facial expressions, which become distorted. You begin to mistake friends for foes.

We made this discovery by doing the following. Participants came into my laboratory and had a full night of sleep. The following morning, we showed them many pictures of a specific individual's face. However, no two pictures were the same. Instead, the facial expression of that one individual varied across the images in a gradient, shifting from friendly (with a slight smile, calming eye aperture, and approachable look) to increasingly stern and threatening (pursed lips, a furrowed brow, and a menacing look in the eyes). Each image of this individual was subtly different from those on either side of it on the emotional gradient, and across tens of pictures, the full range of intent was expressed, from very prosocial (friendly) to strongly antisocial (unfriendly).

Participants viewed the faces in a random fashion while we scanned their brains in an MRI machine, and they rated how approachable or threatening the images were. The MRI scans allowed us to measure how their brains were interpreting and accurately parsing the threatening facial expressions from the friendly ones after having had a full night of sleep. All the participants repeated the same experiment, but this time we deprived them of sleep, including the critical stage of REM. Half of the participants went through the sleep deprivation session first, followed by the sleep session second, and vice versa. In each session, a different individual was featured in the pictures, so there was no memory or repetition effects.

Having had a full night of sleep, which contained REM sleep, participants demonstrated a beautifully precise tuning curve of emotional face recognition, rather like a stretched out V shape. When navigating the cornucopia of facial expressions we showed them inside the MRI scanner, their brains had no problem deftly separating one emotion from another across the delicately changing gradient, and the accuracy of their own ratings proved this to be similarly true. It was effortless to disambiguate friendly and approachable signals from those intimating even minor threat as the emotional tide changed toward the foreboding.

Confirming the importance of the dream state, the better the quality of REM sleep from one individual to the next across that rested night, the more precise the tuning within the emotional decoding networks of the brain the next day. Through this platinum-grade nocturnal service, better REM-sleep quality at night provided superior comprehension of the social world the next day.

But when those same participants were deprived of sleep, including the essential influence of REM sleep, they could no longer distinguish one emotion from another with accuracy. The tuning V of the brain had been changed, rudely pulled all the way up from the base and flattened into a horizontal line, as if the brain was in a state of generalized hypersensitivity without the ability to map gradations of emotional signals from the outside world. Gone was the precise ability to read giveaway clues in another's face. The brain's emotional navigation system had lost its true magnetic north of directionality and sensitivity: a compass that otherwise guides us toward numerous evolutionary advantages.

With the absence of such emotional acuity, normally gifted by the re-tuning skills of REM sleep at night, the sleep-deprived participants slipped into a default of fear bias, believing even gentle- or somewhat friendly looking faces were menacing. The outside world had become a more threatening and aversive place when the brain lacked REM sleep—untruthfully so. Reality and perceived reality were no longer the same in the "eyes" of the sleepless brain. By removing REM sleep, we had, quite literally, removed participants' levelheaded ability to read the social world around them.

Now think of occupations that require individuals to be sleep-deprived, such as law enforcement and military personnel, doctors, nurses, and those in the emergency services—not to mention the ultimate caretaking job: new parents. Every one of these roles demands the accurate ability to read the emotions of others in order to make critical, even life-dependent, decisions, such as detecting a true threat that requires the use of weapons, assessing emotional discomfort or anguish that can change a diagnosis, the extent of palliative pain medication prescribed, or deciding when to express compassion or dispense an assertive parenting lesson. Without REM sleep and its ability to reset the brain's emotional compass, those same individuals will be inaccurate in their social and emotional comprehension of the world around them, leading to inappropriate decisions and actions that may have grave consequences.

Looking across the life span, we have discovered that this REM-sleep recalibration service comes into its own just prior to the transition into adolescence. Before that, when children are still under close watch from their parents, and many salient assessments and decisions are made by Mom and/or Dad, REM sleep provides less of a re-tuning benefit to a child's brain. But come the early teenage years and the inflection point of parental independence wherein an adolescent must navigate the socioemotional world for himself, now we see the young brain feasting on this emotional recalibration benefit of REM sleep. That is *not* to suggest that REM sleep is unnecessary for children or infants—it very much is, as it supports other functions we have discussed (brain development) and will next discuss (creativity). Rather, it is that this particular function of REM sleep, which takes hold at a particular developmental milestone, allows the burgeoning pre-adult brain to steer itself through the turbulent waters of a complex emotional world with autonomy.

We shall return to this topic in the penultimate chapter when we discuss the damage that early school start times are having on our teenagers. Most significant is the issue of sunrise school bus schedules that selectively deprive our teenagers of that early-morning slumber, just at the moment in their sleep cycle when their developing brains are about to drink in most of their much-needed REM sleep. We are bankrupting their dreams, in so many different ways.

CHAPTER 11

Dream Creativity and Dream Control

Aside from being a stoic sentinel that guards your sanity and emotional well-being, REM sleep and the act of dreaming have another distinct benefit: intelligent information processing that inspires creativity and promotes problem solving. So much so, that some individuals try controlling this normally non-volitional process and direct their own dream experiences while dreaming.

DREAMING: THE CREATIVE INCUBATOR

Deep NREM sleep strengthens individual memories, as we now know. But it is REM sleep that offers the masterful and complementary benefit of fusing and blending those elemental ingredients together, in abstract and highly novel ways. During the dreaming sleep state, your brain will cogitate vast swaths of acquired knowledge,* and then extract overarching rules and commonalities—"the gist." We awake with a revised "Mind Wide Web" that is capable of divining solutions to previously impenetrable problems. In this way, REM-sleep dreaming is informational alchemy.

From this dreaming process, which I would describe as ideasthesia, have come some of the most revolutionary leaps forward in human progress. There is perhaps no better illustration highlighting the smarts

*One example is language learning, and the extraction of new grammatical rules. Children exemplify this. They will start using the laws of grammar (e.g., conjunctions, tenses, pronouns, etc.) long before they understand what these things are. It is during sleep that their brains implicitly extract these rules, based on waking experience, despite the child lacking explicit awareness of the rules.

of REM-sleep dreaming than the elegant solution to everything we know of, and how it fits together. I am not trying to be obtuse. Rather, I am describing the dream of Dmitri Mendeleev on February 17, 1869, which led to the periodic table of elements: the sublime ordering of all known constituent building blocks of nature.

Mendeleev, a Russian chemist of renowned ingenuity, had an obsession. He felt there might be an organizational logic to the known elements in the universe, euphemistically described by some as the search for God's abacus. As proof of his obsession, Mendeleev made his own set of playing cards, with each card representing one of the universal elements and its unique chemical and physical properties. He would sit in his office, at home, or on long train rides, and maniacally deal the shuffled deck down onto a table, one card at a time, trying to deduce the rule of all rules that would explain how this ecumenical jigsaw puzzle fit together. For years he pondered the riddle of nature. For years he failed.

After allegedly having not slept for three days and three nights, he'd reached a crescendo of frustration with the challenge. While the extent of sleep deprivation seems unlikely, a clear truth was Mendeleev's continued failure to crack the code. Succumbing to exhaustion, and with the elements still swirling in his mind and refusing organized logic, Mendeleev lay down to sleep. As he slept, he dreamed, and his dreaming brain accomplished what his waking brain was incapable of. The dream took hold of the swirling ingredients in his mind and, in a moment of creative brilliance, snapped them together in a divine grid, with each row (period) and each column (group) having a logical progression of atomic and orbiting electron characteristics, respectively. In Mendeleev's own words:*

> I saw in a dream a table where all the elements fell into place as required. Awakening, I immediately wrote it down on a piece of paper. Only in one place did a correction later seem necessary.

While some contest how complete the dream solution was, no one challenged the evidence that Mendeleev was provided a dream-inspired

*Quoted by B. M. Kedrov in his text, "On the question of the psychology of scientific creativity (on the occassion of the discovery by D. I. Mendeleev of the periodic law)." *Soviet Psychology*, 1957, 3: 91–113.

formulation of the periodic table. It was his dreaming brain, not his waking brain, that was able to perceive an organized arrangement of all known chemical elements. Leave it to REM-sleep dreaming to solve the baffling puzzle of how all constituents of the known universe fit together—an inspired revelation of cosmic magnitude.

My own field of neuroscience has been the beneficiary of similar dream-fueled revelations. The most impactful is that of neuroscientist Otto Loewi. Loewi dreamed of a clever experiment on two frogs' hearts that would ultimately reveal how nerve cells communicate with each other using chemicals (neurotransmitters) released across tiny gaps that separate them (synapses), rather than direct electrical signaling that could only happen if they were physically touching each other. So profound was this dream-implanted discovery that it won Loewi a Nobel Prize.

We also know of precious artistic gifts that have arisen from dreams. Consider Paul McCartney's origination of the songs "Yesterday" and "Let It Be." Both came to McCartney in his sleep. In the case of "Yesterday," McCartney recounts the following dream-inspired awakening while he was staying in a small attic room of his family's house on Wimpole Street, London, during the filming of the delightful movie *Help*:

> I woke up with a lovely tune in my head. I thought, "That's great, I wonder what that is?" There was an upright piano next to me, to the right of the bed by the window. I got out of bed, sat at the piano, found G, found F sharp minor 7th—and that leads you through then to B to E minor, and finally back to E. It all leads forward logically. I liked the melody a lot, but because I'd dreamed it, I couldn't believe I'd written it. I thought, "No, I've never written anything like this before." But I had, which was the most magic thing!

Having been born and raised in Liverpool, I am admittedly biased toward emphasizing the dreaming brilliance of the Beatles. Not to be outdone, however, Keith Richards of the Rolling Stones has arguably the best sleep-inspired story, which gave rise to the opening riff of their song "Satisfaction." Richards would routinely keep a guitar and tape recorder at his bedside to record ideas that would come to him in the night. He describes the following experience on May 7, 1965, after having returned to his hotel room in Clearwater, Florida, following a performance that evening:

> I go to bed as usual with my guitar, and I wake up the next morning, and I see that the tape is run to the very end. And I think, "Well, I didn't do anything. Maybe I hit a button when I was asleep." So I put it back to the beginning and pushed play and there, in some sort of ghostly version, is [the opening lines to "Satisfaction"]. It was a whole verse of it. And after that, there's 40 minutes of me snoring. But there's the song in its embryo, and I actually dreamt the damned thing.

The creative muse of dreaming has also sparked countless literary ideas and epics. Take the author Mary Shelley, who passed through a most frightening dream scene one summer night in 1816 while staying in one of Lord Byron's estates near Lake Geneva—a dream she almost took to be waking reality. That dreamscape gave Shelley the vision and narrative for the spectacular gothic novel *Frankenstein*. Then there is the French surrealist poet St. Paul Boux, who well understood the fertile talents of dreaming. Before retiring each night, he is said to have hung a sign on his bedroom door that read: "Do Not Disturb: Poet at Work."*

Anecdotes such as these are enjoyable stories to tell, but they do not serve as experimental data. What, then, is the scientific evidence establishing that sleep, and specifically REM sleep and dreaming, provides a form of associative memory processing—one that fosters problem solving? And what is so special about the neurophysiology of REM sleep that would explain these creative benefits, and the dreaming obligate to them?

REM-SLEEP FUZZY LOGIC

An obvious challenge to testing the brain when it is asleep is that . . . it is asleep. Sleeping individuals cannot engage in computerized tests

---

*This ode to the creative juices of dream sleep is sometimes also attributed to the French Symbolist poet Paul-Pierre Roux.

nor provide useful responses—the typical way that cognitive scientists assess the workings of the brain. Short of lucid dreaming, which we will address at the end of this chapter, sleep scientists have been left wanting in this regard. We have frequently been resigned to passively observing brain activity during sleep, without ever being able to have participants perform tests while they are sleeping. Rather, we measure waking performance before and after sleep and determine if the sleep stages or dreaming that occurred in between explains any observed benefit the next day.

I and my colleague at Harvard Medical School Robert Stickgold designed a solution to this problem, albeit an indirect and imperfect one. In chapter 7 I described the phenomenon of sleep inertia—the carryover of the prior sleeping brain state into wakefulness in the minutes after waking up. We wondered whether we could turn this brief window of sleep inertia to our experimental advantage—not by waking subjects up in the morning and testing them, but rather by waking individuals up from different stages of NREM sleep and REM sleep throughout the night.

The dramatic alterations in brain activity during NREM and REM sleep, and their tidal shifts in neurochemical concentrations, do not reverse instantaneously when you awaken. Instead, the neural and chemical properties of that particular sleep stage will linger, creating the inertia period that separates true wakefulness from sleep, and last some minutes. Upon enforced awakening, the brain's neurophysiology starts out far more sleep-like than wake-like and, with each passing minute, the concentration of the prior sleep stage from which an individual has been woken will gradually fade from the brain as true wakefulness rises to the surface.

By restricting the length of whatever cognitive test we performed to just ninety seconds, we felt we could wake individuals up and very quickly test them in this transitional sleep phase. In doing so, we could perhaps capture some of the functional properties of the sleep stage from which the participant was woken, like capturing the vapors of an evaporating substance and analyzing those vapors to draw conclusions about the properties of the substance itself.

It worked. We developed an anagram task in which the letters of real words were scrambled. Each word was composed of five letters, and the anagram puzzles only had one correct solution (e.g., "OSEOG" = "GOOSE"). Participants would see the scrambled words one at a time on the screen for just a few seconds, and they were asked to speak the solution, if they had one, before the time ran out and the next anagram word puzzle appeared on the screen. Each test session lasted only ninety seconds, and we recorded how many problems the participants correctly solved within this brief inertia period. We would then let the participants fall back asleep.

The subjects had the task described to them before going to bed in the sleep laboratory with electrodes placed on the head and face so that I could measure their sleep unfolding in real time on a monitor next door. The participants also performed a number of trials before getting into bed, allowing them to get familiar with the task and how it worked. After falling asleep, I then woke subjects up four times throughout the night, twice from NREM sleep early and late in the night, and twice from REM sleep, also early and late in the night.

Upon awakenings from NREM sleep, participants did not appear to be especially creative, solving few of the anagram puzzles. But it was a different story when I woke them up out of REM sleep, from the dreaming phase. Overall, problem-solving abilities rocketed up, with participants solving 15 to 35 percent more puzzles when emerging from REM sleep compared with awakenings from NREM sleep or during daytime waking performance!

Moreover, the way in which the participants were solving the problems after exiting REM sleep was different from how they solved the problems both when emerging from NREM sleep and while awake during the day. The solutions simply "popped out" following awakenings from REM sleep, one subject told me, though at the time, they did not know they had been in REM sleep just prior. Solutions seemed more effortless when the brain was being bathed by the afterglow of dream sleep. Based on response times, solutions arrived more instantaneously following an REM sleep awakening, relative to the slower, deliberative solutions that came when that same individual was exiting NREM sleep or when they were awake during the day. The lingering vapors of REM sleep were providing a more fluid, divergent, "open-minded" state of information processing.

Using the same type of experimental awakening method, Stickgold performed another clever test that reaffirmed how radically different the REM-sleep dreaming brain operates when it comes to creative memory processing. He examined the way in which our stores of related concepts, also known as semantic knowledge, function at night. It's this semantic knowledge like a pyramidal family tree of relatedness that fans out from top to bottom in order of relatedness strength. Figure 14 is an example of one such associative web plucked from my own mind regarding UC Berkeley, where I am a professor:

Figure 14: Example of a Memory Association Network

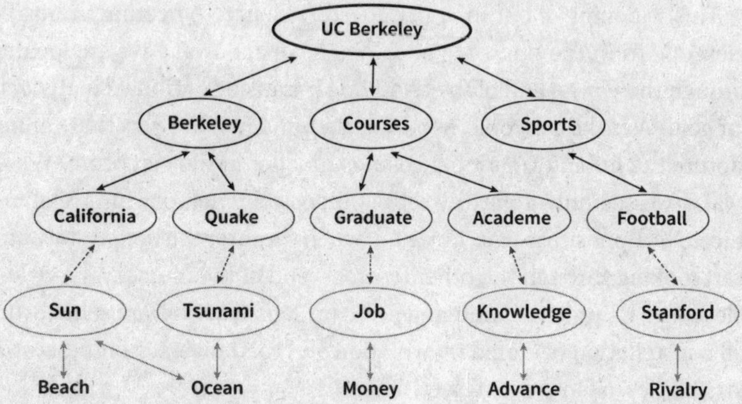

Using a standard computer test, Stickgold measured how these associative networks of information operated following NREM-sleep and REM-sleep awakenings, and during standard performance during the waking day. When you wake the brain from NREM or measure performance during the day, the operating principles of the brain are closely and logically connected, just as pictured in figure 14. However, wake the brain up from REM sleep and the operating algorithm was completely different. Gone is the hierarchy of logical associative connection. The REM-sleep dreaming brain was utterly uninterested in bland, commonsense links—the one-step-to-the-next associations. Instead, the REM-sleep brain was shortcutting the obvious links and favoring very distantly related concepts. The logic guards had left the REM-sleep dreaming brain. Wonderfully eclectic lunatics were now running the associative memory asylum. From the REM-sleep dreaming state, almost anything goes—and the more bizarre the better, the results suggested.

The two experiments of anagram solving and semantic priming revealed how radically different the operating principles of the dreaming brain were, relative to those of NREM sleep and wakefulness. As we enter REM sleep and dreaming takes hold, an inspired form of memory mixology begins to occur. No longer are we constrained to see the most typical and plainly obvious connections between memory units. On the contrary, the brain becomes actively biased toward seeking out the most distant, nonobvious links between sets of information.

This widening of our memory aperture is akin to peering through a telescope from the opposing end. When we are awake we are looking through the wrong end of the telescope if transformational creativity is our goal. We take a myopic, hyperfocused, and narrow view that cannot capture the full informational cosmos on offer in the cerebrum. When awake, we see only a narrow set of all possible memory interrelationships. The opposite is true, however, when we enter the dream state and start looking through the other (correct) end of the memory-surveying telescope. Using that wide-angle dream lens, we can apprehend the full constellation of stored information and their diverse combinatorial possibilities, all in creative servitude.

MEMORY MELDING IN THE FURNACE OF DREAMS

Overlay these two experimental findings onto the dream-inspired-problem-solving claims, such as those of Dmitri Mendeleev, and two clear, scientifically testable hypotheses emerge.

First, if we feed a waking brain with the individual ingredients of a problem, novel connections and problem solutions should preferentially—if not exclusively—emerge after time spent in the REM dreaming state, relative to an equivalent amount of deliberative time spent awake. Second, the content of people's dreams, above and beyond simply having REM sleep, should determine the success of those hyper-associative problem-solving benefits. As with the effects of REM sleep on our emotional and mental well-being explored in the previous chapter, the latter would prove that REM sleep is necessary but not sufficient. It is both the act of dreaming and the associated content of those dreams that determine creative success.

That is precisely what we and others have found time and again. As an example, let's say that I teach you a simple relationship between two objects, A and B, such that A should be chosen over object B (A>B). Then I teach you another relationship, which is that object B should be chosen over object C (B>C). Two separate, isolated premises. If I then show you A and C together, and ask you which you would choose, you would very likely pick A over C because your brain made an inferential leap. You took two preexisting memories (A>B and B>C) and, by flexibly interrelating them (A>B>C), came up with a completely novel answer to a previously unasked question (A>C). This is the power of relational memory processing, and it is one that receives an accelerated boost from REM sleep.

In a study conducted with my Harvard colleague Dr. Jeffrey Ellenbogen, we taught participants lots of these individual premises that were nested in a large chain of interconnectedness. Then we gave them tests that assessed not just their knowledge of these individual pairs, but also assessed whether they knew how these items connected together in the associative chain. Only those who had slept and obtained late-morning REM sleep, rich in dreaming, showed evidence of linking the memory elements together (A>B>C>D>E>F, etc.), making them capable of the most distant associative leaps (e.g., B>E). The very same benefit was found after daytime naps of sixty to ninety minutes that also included REM sleep.

It is sleep that builds connections between distantly related informational elements that are not obvious in the light of the waking day. Our participants went to bed with disparate pieces of the jigsaw and woke up with the puzzle complete. It is the difference between knowledge (retention of individual facts) and wisdom (knowing what they all mean when you fit them together). Or, said more simply, learning versus comprehension. REM sleep allows your brain to move beyond the former and truly grasp the latter.

Some may consider this informational daisy-chaining to be trivial, but it is one of the key operations differentiating your brain from your computer. Computers can store thousands of individual files with precision. But standard computers do not intelligently interlink those files in numerous and creative combinations. Instead, computer files sit like isolated islands. Our human memories are, on the other hand, richly interconnected in webs of associations that lead to flexible, predictive powers. We have REM sleep, and the act of dreaming, to thank for much of that inventive hard work.

CODE CRACKING AND PROBLEM SOLVING

More than simply melding information together in creative ways, REM-sleep dreaming can take things a step further. REM sleep is capable of creating *abstract* overarching knowledge and super-ordinate concepts out of sets of information. Think of an experienced physician who is able to seemingly intuit a diagnosis from the many tens of varied, subtle symptoms she observes in a patient. While this kind of abstractive skill can come after years of hard-earned experience, it is also the very same accurate gist extraction that we have observed REM sleep accomplishing within just one night.

A delightful example is observed in infants abstracting complex grammatical rules in a language they must learn. Even eighteen-month-old babies have been shown to deduce high-level grammatical structure from novel languages they hear, but only after they have slept following the initial exposure. As you will recall, REM sleep is especially dominant during this early-life window, and it is that REM sleep that plays a critical role in the development of language, we believe. But that benefit extends beyond infancy—very similar results have been reported in adults who are required to learn new language and grammar structures.

Perhaps the most striking proof of sleep-inspired insight, and one I most frequently describe when giving talks to start-up, tech, or innovative business companies to help them prioritize employee sleep, comes from a study conducted by Dr. Ullrich Wagner at the University of Lübeck, Germany. Trust me when I say you'd really rather not be a participant in these experiments. Not because you have to suffer extreme sleep deprivation for days, but because you have to work through hundreds of miserably laborious number-string problems, almost like having to do long division for an hour or more. Actually "laborious" is far too generous a description. It's possible some people have lost the will to live while trying to sit and solve hundreds of these number problems! I know, I've taken the test myself.

You will be told that you can work through these problems using specific rules that are provided at the start of the experiment. Sneakily, what the researchers do not tell you about is the existence of a hidden rule, or shortcut, common across all the problems. If you figure out this embedded cheat, you can solve many more problems in a far shorter time. I'll return to this shortcut in just a minute. After having had participants perform hundreds of these problems, they were to return twelve hours later and once again work through hundreds more of these mind-numbing problems. However, at the end of this second test session, the researchers asked whether the subjects had cottoned on to the hidden rule. Some of the participants spent that twelve-hour time delay awake across the day, while for others, that time window included a full eight-hour night of sleep.

After time spent awake across the day, despite the chance to consciously deliberate on the problem as much as they desired, a rather paltry 20 percent of participants were able to extract the embedded shortcut. Things were very different for those participants who had obtained a full night of sleep—one dressed with late-morning, REM-rich slumber. Almost 60 percent returned and had the "ah-ha!" moment of spotting the hidden cheat—which is a threefold difference in creative solution insight afforded by sleep!

Little wonder, then, that you have never been told to "stay awake on a problem." Instead, you are instructed to "sleep on it." Interestingly, this phrase, or something close to it, exists in most languages (from the French *dormir sur un problem*, to the Swahili *kulala juu ya tatizo*), indicating that the problem-solving benefit of dream sleep is universal, common across the globe.

FUNCTION FOLLOWS FORM — DREAM CONTENT MATTERS

The author John Steinbeck wrote, "A problem difficult at night is resolved in the morning after the committee of sleep has worked on it." Should he have prefaced "committee" with the word "dream"? It appears so. The *content* of one's dreams, more than simply dreaming per se, or even sleeping, determines problem-solving success. Though such a claim has long been made, it took the advent of virtual reality for us to prove as much—and in the process, shore up the claims of Mendeleev, Loewi, and many other nocturnal troubleshooters.

Enter my collaborator Robert Stickgold, who designed a clever experiment in which participants explored a computerized virtual reality maze. During an initial learning session, he would start participants off from different random locations within the virtual maze and ask them to navigate their way out through exploratory trial and error. To aid their learning, Stickgold placed unique objects, such as a Christmas tree, to act as orientation or anchor points at specific locations within the virtual maze.

Almost a hundred research participants explored the maze during the first learning session. Thereafter, half of them took a ninety-minute nap, while the other half remained awake and watched a video, all monitored with electrodes placed on the head and face. Throughout the ninety-minute epoch, Stickgold would occasionally wake the napping individuals and ask them about the content of any dreams they were having, or for the group that remained awake, ask them to report any particular thoughts that were going through their minds at the time. Following the ninety-minute period, and after another hour or so to overcome sleep inertia in those who napped, everyone was dropped back into the virtual maze and tested once more to see if their performance was any better than during initial learning.

It should come as no surprise by now that those participants who took a nap showed superior memory performance on the maze task. They could locate the navigation clues with ease, finding their way around and out of the maze faster than those who had not slept. The novel result, however, was the difference that dreaming made. Participants who slept and reported dreaming of elements of the maze, and themes around experiences clearly related to it, showed almost ten times more improvement in their task performance upon awakening than those who slept just as much, and also dreamed, but did not dream of maze-related experiences.

As in his earlier studies, Stickgold found that the dreams of these super-navigators were not a precise replay of the initial learning experience while awake. For example, one participant's dream report stated: "I was thinking about the maze and kinda having people as checkpoints, I guess, and then that led me to think about when I went on this trip a few years ago and we went to see these bat caves, and they're kind of like, maze-like." There were no bats in Stickgold's virtual maze, nor were there any other people or checkpoints. Clearly, the dreaming brain was not simply recapitulating or re-creating exactly what happened to them in the maze. Rather, the dream algorithm was cherry-picking salient fragments of the prior learning experience, and then attempting to place those new experiences within the back catalog of preexisting knowledge.

Like an insightful interviewer, dreaming takes the approach of interrogating our recent autobiographical experience and skillfully positioning it within the context of past experiences and accomplishments, building a rich tapestry of meaning. "How can I understand and connect that which I have recently learned with that I already know, and in doing so, discover insightful new links and revelations?" Moreover, "What have I done in the past that might be useful in potentially solving this newly experienced problem in the future?" Different from solidifying memories, which we now realize to be the job of NREM sleep, REM sleep, and the act of dreaming, takes that which we have learned in one experience setting and seeks to apply it to others stored in memory.

When I have discussed these scientific discoveries in public lectures, some individuals will question their validity on the grounds of historical legends who were acclaimed short-sleepers, yet still demonstrated remarkable creative prowess. One common name that I frequently encounter in such rebuttals is the inventor Thomas Edison. We will never truly know if Edison was the short-sleeper that some, including himself, claim. What we do know, however, is that Edison was a habitual daytime napper. He understood the creative brilliance of dreaming, and used it ruthlessly as a tool, describing it as "the genius gap."

Edison would allegedly position a chair with armrests at the side of his study desk, on top of which he would place a pad of paper and a pen. Then he would take a metal saucepan and turn it upside down, carefully positioning it on the floor directly below the right-side armrest of the chair. If that were not strange enough, he would pick up two or three steel ball bearings in his right hand. Finally, Edison would settle himself down into the chair, right hand supported by the armrest, grasping the ball bearings. Only then would Edison ease back and allow sleep to consume him whole. At the moment he began to dream, his muscle tone would relax and he would release the ball bearings, which would crash on the metal saucepan below, waking him up. He would then write down all of the creative ideas that were flooding his dreaming mind. Genius, wouldn't you agree?

CONTROLLING YOUR DREAMS—LUCIDITY

No chapter on dreaming can go unfinished without mention of lucidity. Lucid dreaming occurs at the moment when an individual becomes aware that he or she is dreaming. However, the term is more colloquially used to describe gaining volitional control of *what* an individual is dreaming, and the ability to manipulate that experience, such as deciding to fly, or perhaps even the functions of it, such as problem solving.

The concept of lucid dreaming was once considered a sham. Scientists debated its very existence. You can understand the skepticism. First, the assertion of conscious control over a normally non-volitional process injects a heavy dose of ludicrous into the already preposterous experience we call dreaming. Second, how can you objectively prove a subjective claim, especially when the individual is fast asleep during the act?

Four years ago, an ingenious experiment removed all such doubt. Scientists placed lucid dreamers inside an MRI scanner. While awake, these participants first clenched their left and then right hand, over and over. Researchers took snapshots of brain activity, allowing them to define the precise brain areas controlling each hand of each individual.

The participants were allowed to fall asleep in the MRI scanner, entering REM sleep where they could dream. During REM sleep, however, all voluntary muscles are paralyzed, preventing the dreamer from acting out ongoing mental experience. Yet, the muscles that control the eyes are spared from this paralysis, and give this stage of sleep its frenetic name. Lucid dreamers were able to take advantage of this ocular freedom, communicating with the researchers through eye movements. Pre-defined eye movements would therefore inform the researchers of the nature of the lucid dream (e.g., the participant made three deliberate leftward eye movements when they gained lucid dream control, two rightward eye movements before clenching their right hand, etc.). Non-lucid dreamers find it difficult to believe that such deliberate eye movements are possible while someone is asleep, but watch a lucid dreamer do it a number of times, and it is impossible to deny.

When participants signaled the beginning of the lucid dream state, the scientists began taking MRI pictures of brain activity. Soon after, the sleeping participants signaled their intent to dream about moving their left hand, then their right hand, alternating over and over again, just as they did when awake. Their hands were not physically moving—they could not, due to the REM-sleep paralysis. But they were moving in the dream.

At least, that was the subjective claim from the participants upon awakening. The results of the MRI scans objectively proved they were not lying. The same regions of the brain that were active during physical right and left voluntary hand movements observed while the individuals were awake similarly lit up in corresponding ways during times when the lucid participants signaled that they were clenching their hands while dreaming!

There could be no question. Scientists had gained objective, brain-based proof that lucid dreamers can control when and what they dream while they are dreaming. Other studies using similar eye movement communication designs have further shown that individuals can deliberately bring themselves to timed orgasm during lucid dreaming, an outcome that, especially in males, can be objectively verified using physiological measures by (brave) scientists.

It remains unclear whether lucid dreaming is beneficial or detrimental, since well over 80 percent of the general populace are not natural lucid dreamers. If gaining voluntary dream control were so useful, surely Mother Nature would have imbued the masses with such a skill.

However, this argument makes the erroneous assumption that we have stopped evolving. It is possible that lucid dreamers represent the next iteration in Homo sapiens' evolution. Will these individuals be preferentially selected for in the future, in part on the basis of this unusual dreaming ability—one that may allow them to turn the creative problem-solving spotlight of dreaming on the waking challenges faced by themselves or the human race, and advantageously harness its power more deliberately?

PART 4

From Sleeping Pills to Society Transformed

CHAPTER 12

Things That Go Bump in the Night

Sleep Disorders and Death Caused by No Sleep

Few other areas of medicine offer a more disturbing or astonishing array of disorders than those concerning sleep. Considering how tragic and remarkable disorders in those other fields can be, this is quite a claim. Yet when you consider that oddities of slumber include daytime sleep attacks and body paralysis, homicidal sleepwalking, dream enactment, and perceived alien abductions, the assertion starts to sound more valid. Most astonishing of all, perhaps, is a rare form of severe insomnia, with the disorder ultimately leading to death, supported by the life-extinguishing upshot of extreme total sleep deprivation in animal studies.

This chapter is by no means a comprehensive review of all sleep disorders, of which there are now over one hundred known. Nor is it meant to serve as a medical guide to any one disorder, since I am not a board certified doctor of sleep medicine, but rather a sleep scientist. For those seeking advice on sleep disorders, I recommend visiting the National Sleep Foundation website,* and there you will find resources on sleep centers near you.

Rather than attempting a quick-fire laundry list of the many tens of sleep disorders that exist, I have chosen to focus on a select few—namely somnambulism, insomnia, narcolepsy, and fatal familial insomnia—from the vantage point of science, and what the science of these disorders can meaningfully teach us about the mysteries of sleeping and dreaming.

---

*https://sleepfoundation.org.

SOMNAMBULISM

The term "somnambulism" refers to sleep (*somnus*) disorders that involve some form of movement (*ambulation*). It encompasses conditions such as sleepwalking, sleep talking, sleep eating, sleep texting, sleep sex, and, very rarely, sleep homicide.

Understandably, most people believe these events happen during REM sleep as an individual is dreaming, and specifically acting out ongoing dreams. However, all these events arise from the deepest stage of non-dreaming (NREM) sleep, and not dream (REM) sleep. If you rouse an individual from a sleepwalking event and ask what was going through their mind, rarely will they report a thing—no dream scenario, no mental experience.

While we do not yet fully understand the cause of somnambulism episodes, the existing evidence suggests that an unexpected spike in nervous system activity during deep sleep is one trigger. This electrical jolt compels the brain to rocket from the basement of deep NREM sleep all the way to the penthouse of wakefulness, but it gets stuck somewhere in between (the thirteenth floor, if you will). Trapped between the two worlds of deep sleep and wakefulness, the individual is confined to a state of mixed consciousness—neither awake nor asleep. In this confused condition, the brain performs basic but well-rehearsed actions, such as walking over to a closet and opening it, placing a glass of water to the lips, or uttering a few words or sentences.

A full diagnosis of somnambulism can require the patient to spend a night or two in a clinical sleep laboratory. Electrodes are placed on the head and body to measure the stages of sleep, and an infrared video camera on the ceiling records the nighttime events, like a single night-vision goggle. At the moment when a sleepwalking event occurs, the video camera footage and the electrical brainwave readouts stop agreeing. One suggests that the other is lying. Watching the video, the patient is clearly "awake" and behaving. They may sit up on the edge of the bed and begin talking. Others may attempt to put on clothes and walk out of the room. But look at the brainwave activity and you realize that the patient, or at least their brain, is sound asleep. There are the clear and unmistakable slow electrical waves of deep NREM sleep, with no sign of fast, frenetic waking brainwave activity.

For the most part, there is nothing pathological about sleepwalking or sleep talking. They are common in the adult population, and even more common in children. It is not clear why children experience somnambulism more than adults, nor is it clear why some children grow out of having these nighttime events, while others will continue to do so throughout their lives. One explanation of the former is simply the fact that we have greater amounts of deep NREM sleep when we are young, and therefore the statistical likelihood of sleepwalking and sleep talking episodes occurring is higher.

Most episodes of the condition are harmless. Occasionally, however, adult somnambulism can result in a much more extreme set of behaviors, such as those performed by Kenneth Parks in 1987. Parks, who was twenty-three years old at the time, lived with his wife and five-month-old daughter in Toronto. He had been suffering from severe insomnia caused by the stress of joblessness and gambling debts. By all accounts, Parks was a nonviolent man. His mother-in-law—with whom he had a good relationship—called him a "gentle giant" on the basis of his placid nature yet considerable height and broad-shouldered form (he stood six foot four, and weighed 225 pounds). Then came May 23.

After falling asleep on the couch around 1:30 a.m. while watching television, Parks arose and got in his car, barefoot. Depending on the route, it is estimated that Parks drove approximately fourteen miles to his in-laws' home. Upon entering the house, Parks made his way upstairs, stabbed his mother-in-law to death with a knife he had taken from their kitchen, and strangled his father-in-law unconscious after similarly attacking him with a cleaver (his father-in-law survived). Parks then got back in his car and, upon regaining full waking consciousness at some point, drove to a police station and said, "I think I have killed some people . . . my hands." Only then did he realize the blood flowing down his arms as a result of severing his own flexor tendons with the knife.

Since he could remember only vague fragments of the murder (e.g., flashes of his mother-in-law's face with a "help me" look on it), had no motive, and had a long history of sleepwalking (as did other members of his family), a team of defense experts concluded that Ken Parks was asleep when he committed the crime, suffering a severe episode of sleepwalking. They argued that he was unaware of his actions, and thus not culpable. On May 25, 1988, a jury rendered a verdict of not guilty. This defense has been attempted in a number of subsequent cases, most of which have been unsuccessful.

The story of Ken Parks is of the most tragic kind, and to this day Parks struggles with a guilt one suspects may never leave him. I offer the account not to scare the reader, nor to try to sensationalize the dire events of that late May night in 1987. Rather, I offer it to illustrate how non-volitional acts arising from sleep and its disorders can have very real legal, personal, and societal consequences, and demand the contribution of scientists and doctors in arriving at the appropriate legal justice.

I also want to note, for the concerned sleepwalkers reading this chapter, that most somnambulism episodes (e.g., sleep walking, talking) are considered benign and do not require intervention. Medicine will usually step in with treatment solutions only if the afflicted patient or his caretaker, partner, or parent (in the case of children) feels that the condition is compromising health or poses a risk. There are effective treatments, and it is a shame one never arrived in time for Ken Parks prior to that ill-fated evening in May.

INSOMNIA

For many individuals these days, shudder quotes have come home to roost around the phrase "a good night's sleep," as the writer Will Self has lamented. Insomnia, to which his grumblings owe their origin, is the most common sleep disorder. Many individuals suffer from insomnia, yet some believe they have the disorder when they do not. Before describing the features and causes of insomnia (and in the next chapter, potential treatment options), let me first describe what insomnia is not—and in doing so, reveal what it is. As someone who has experienced his own bouts of insomnia in life, I can certainly sympathize with those who are struggling with this condition.

Being sleep deprived is not insomnia. In the field of medicine, sleep deprivation is considered as (i) having the *adequate ability* to sleep; yet

(ii) giving oneself an *inadequate opportunity* to sleep—that is, sleep-deprived individuals can sleep, if only they would take the appropriate time to do so. Insomnia is the opposite: (i) suffering from an *inadequate ability* to generate good sleep quality or quantity, despite (ii) allowing oneself the *adequate opportunity* to get sleep. People suffering from insomnia therefore cannot produce sufficient sleep quantity/quality, even though they give themselves enough time to do so (seven to nine hours).

Before moving on, it is worth noting the condition of sleep-state misperception, also known as paradoxical insomnia. Here, patients will report having slept poorly throughout the night, or even not sleeping at all. However, when these individuals have their sleep monitored objectively using electrodes or other accurate sleep monitoring devices, there is a mismatch. The sleep recordings indicate that the patient has slept far better than they themselves believe, and sometimes indicate that a completely full and healthy night of sleep occurred. Patients suffering from paradoxical insomnia therefore have an illusion, or misperception, of poor sleep that is not actually poor. As a result, such patients are treated as hypochondriacal. Though the term may seem dismissive or condescending, it is taken very seriously by sleep medicine doctors, and there are psychological interventions that help after the diagnosis is made.

Returning to the condition of true insomnia, there are several different sub-types, in the same way that there are numerous different forms of cancer, for example. One distinction separates insomnia into two kinds. The first is sleep *onset* insomnia, which is difficulty falling asleep. The second is sleep *maintenance* insomnia, or difficulty staying asleep. As the actor and comedian Billy Crystal has said when describing his own battles with insomnia, "I sleep like a baby—I wake up every hour." Sleep onset and sleep maintenance insomnia are not mutually exclusive: you can have one or the other, or both. No matter which of these kinds of sleep problems is occurring, sleep medicine has very specific clinical boxes that must be checked for a patient to receive a diagnosis of insomnia. For now, these are:

- Dissatisfaction with sleep quantity or quality (e.g., difficulty falling asleep, staying sleep, early-morning awakening)

- Suffering significant distress or daytime impairment
- Has insomnia at least three nights each week for more than three months
- Does not have any coexisting mental disorders or medical conditions that could otherwise cause what appears to be insomnia

What this really means in terms of boots-on-the-ground patient descriptions is the following chronic situation: difficulty falling asleep, waking up in the middle of the night, waking up too early in the morning, difficulty falling back to sleep after waking up, and feeling unrefreshed throughout the waking day. If any of the characteristics of insomnia feel familiar to you, and have been present for *several months*, I suggest you consider seeking out a sleep medicine doctor. I emphasize a sleep medicine doctor and not necessarily your GP, since GPs—superb as they often are—have surprisingly minimal sleep training during the entirety of medical school and residency. Some GPs are understandably apt to prescribe a sleeping pill, which is rarely the right answer, as we will see in the next chapter.

The emphasis on duration of the sleep problem (more than three nights a week, for more than three months) is important. All of us will experience difficulty sleeping every now and then, which may last just one night or several. That is normal. There is usually an obvious cause, such as work stress or a flare-up in a social or romantic relationship. Once these things subside, though, the sleep difficulty usually goes away. Such acute sleep problems are generally not recognized as chronic insomnia, since clinical insomnia requires an ongoing duration of sleep difficulty, week after week after week.

Even with this strict definition, chronic insomnia is disarmingly common. Approximately one out of every nine people you pass on the street will meet the strict clinical criteria for insomnia, which translates to more than 40 million Americans struggling to make it through their waking days due to wide-eyed nights. While the reasons remain unclear, insomnia is almost twice as common in women than in men, and it is unlikely that a simple unwillingness of men to admit sleep problems explains this very sizable difference between the two sexes. Race and ethnicity also make a significant difference, with African Americans and Hispanic Americans suffering higher rates of insomnia than Caucasian Americans—findings that have important implications for well-recognized health disparities in these communities, such as diabetes, obesity, and cardiovascular disease, which have known links to a lack of sleep.

In truth, insomnia is likely to be a more widespread and serious problem than even these sizable numbers suggest. Should you relax the stringent clinical criteria and just use epidemiological data as a guide, it is probable that two out of every three people reading this book will regularly have difficulty falling or staying asleep at least one night a week, every week.

Without belaboring the point, insomnia is one of the most pressing and prevalent medical issues facing modern society, yet few speak of it this way, recognize the burden, or feel there is a need to act. That the "sleep aid" industry, encompassing prescription sleeping medications and over-the-counter sleep remedies, is worth an astonishing $30 billion a year in the US is perhaps the only statistic one needs in order to realize how truly grave the problem is. Desperate millions of us are willing to pay a lot of money for a good night's sleep.

But dollar values do not address the more important issue of what's causing insomnia. Genetics plays a role, though it is not the full answer. Insomnia shows some degree of genetic heritability, with estimates of 28 to 45 percent transmission rates from parent to child. However, this still leaves the majority of insomnia being associated with non-genetic causes, or gene-environment (nature-nurture) interactions.

To date, we have discovered numerous triggers that cause sleep difficulties, including psychological, physical, medical, and environmental factors (with aging being another, as we have previously discussed). External factors that cause poor sleep, such as too much bright light at night, the wrong ambient room temperature, caffeine, tobacco, and alcohol consumption—all of which we'll visit in more detail in the next chapter—can masquerade as insomnia. However, their origins are not from *within* you, and therefore not a disorder *of* you. Rather, they are influences from outside and, once they are addressed, individuals will get better sleep, without changing anything about themselves.

Other factors, however, come from within a person, and are innate biological causes of insomnia. Noted in the clinical criteria described above, these factors cannot be a symptom of a disease (e.g., Parkinson's disease) or a side effect of a medication (e.g., asthma medication). Rather, the cause(s) of the sleep problem must stand alone in order for you to be primarily suffering from true insomnia.

The two most common triggers of chronic insomnia are psychological: (1) emotional concerns, or worry, and (2) emotional distress, or anxiety. In this fast-paced, information-overloaded modern world, one of the few times that we stop our persistent informational consumption and inwardly reflect is when our heads hit the pillow. There is no worse time to consciously do this. Little wonder that sleep becomes nearly impossible to initiate or maintain when the spinning cogs of our emotional minds start churning, anxiously worrying about things we did today, things that we forgot to do, things that we must face in the coming days, and even those far in the future. That is no kind of invitation for beckoning the calm brainwaves of sleep into your brain, peacefully allowing you to drift off into a full night of restful slumber.

Since psychological distress is a principal instigator of insomnia, researchers have focused on examining the biological causes that underlie emotional turmoil. One common culprit has become clear: an overactive sympathetic nervous system, which, as we have discussed in previous chapters, is the body's aggravating fight-or-flight mechanism. The sympathetic nervous system switches on in response to threat and acute stress that, in our evolutionary past, was required to mobilize a legitimate fight-or-flight response. The physiological consequences are increased heart rate, blood flow, metabolic rate, the release of stress-negotiating chemicals such as cortisol, and increased brain activation, all of which are beneficial in the acute moment of true threat or danger. However, the fight-or-flight response is not meant to be left in the "on" position for any prolonged period of time. As we have already touched upon in earlier chapters, chronic activation of the fight-or-flight nervous system causes myriad health problems, one of which is now recognized to be insomnia.

Why an overactive fight-or-flight nervous system prevents good sleep can be explained by several of the topics we have discussed so far, and some we have not. First, the raised metabolic rate triggered by fight-or-flight nervous system activity, which is common in insomnia patients, results in a higher core body temperature. You may remember from chapter 2 that we must drop core body temperature to initiate and maintain sleep, which becomes more difficult in insomnia patients suffering a raised metabolic rate and higher operating internal temperature, including in the brain.

Second are higher levels of the alertness-promoting hormone cortisol, and sister neurochemicals adrenaline and noradrenaline. All three of these chemicals raise heart rate. Normally, our cardiovascular system calms down as we make the transition into light and then deep sleep. Elevated cardiac activity makes that transition more difficult. All three of these chemicals increase metabolic rate, additionally increasing core body temperature, which further compounds the first problem outlined above.

Third, and related to these chemicals, are altered patterns of brain activity linked with the body's sympathetic nervous system. Researchers have placed healthy sleepers and insomnia patients in a brain scanner and measured the changing patterns of activity as both groups try to fall asleep. In the good sleepers, the parts of the brain related to inciting emotions (the amygdala) and those linked to memory retrospection (the hippocampus) quickly ramped down in their levels of activity as they transitioned toward sleep, as did basic alertness regions in the brain stem. This was not the case for the insomnia patients. Their emotion-generating regions and memory-recollection centers all remained active. This was similarly true of the basic vigilance centers in the brain stem that stubbornly continued their wakeful watch. All the while the thalamus—the sensory gate of the brain that needs to close shut to allow sleep—remained active and open for business in insomnia patients.

Simply put, the insomnia patients could not disengage from a pattern of altering, worrisome, ruminative brain activity. Think of a time when you closed the lid of a laptop to put it to sleep, but came back later to find that the screen was still on, the cooling fans were still running, and the computer was still active, despite the closed lid. Normally this is because programs and routines are still running, and the computer cannot make the transition into sleep mode.

Based on the results of brain-imaging studies, an analogous problem is occurring in insomnia patients. Recursive loops of emotional programs, together with retrospective and prospective memory loops, keep playing in the mind, preventing the brain from shutting down and switching into sleep mode. It is telling that a direct and causal connection exists between the fight-or-flight branch of the nervous system and all of these emotion-, memory-, and alertness-related regions of the brain. The bidirectional line of communication between the body and brain amounts to a vicious, recurring cycle that fuels their thwarting of sleep.

The fourth and final set of identified changes has been observed in the quality of sleep of insomnia patients when they do finally drift off. Once again, these appear to have their origins in an overactive fight-or-flight nervous system. Patients with insomnia have a lower quality of sleep, reflected in shallower, less powerful electrical brainwaves during deep NREM. They also have more fragmented REM sleep, peppered by brief awakenings that they are not always aware of, yet still cause a degraded quality of dream sleep. All of which means that insomnia patients can wake up not feeling refreshed. Consequentially, patients are unable to function well during the day, cognitively and/or emotionally. In this way, insomnia is really a 24/7 disorder: as much a disorder of the day as of the night.

You can now understand how physiologically complex the underlying condition is. No wonder the blunt instruments of sleeping pills, which can sedate your higher brain, or cortex, are no longer recommended as the first-line treatment approach for insomnia by the American Medical Association. Fortunately, a non-pharmacological therapy, which we will discuss in detail in the next chapter, has been developed. It is more powerful in restoring naturalistic sleep in insomnia sufferers, and it elegantly targets each of the physiological components of insomnia described above. Real optimism is to be found in these new, non-drug therapies that I urge you to explore should you suffer from insomnia.

NARCOLEPSY

I suspect that you cannot recall any truly significant action in your life that wasn't governed by two very simple rules: staying away from something that would feel bad, or trying to accomplish something that would feel good. This law of approach and avoidance dictates most of human and animal behavior from a very early age.

The forces that implement this law are positive and negative emotions. Emotions make us do things, as the name suggests (remove the first letter from the word). They motivate our remarkable achievements, incite us to try again when we fail, keep us safe from potential harm, urge us to accomplish rewarding and beneficial outcomes, and compel us to cultivate social and romantic relationships. In short, emotions in appropriate amounts make life worth living. They offer a healthy and vital existence, psychologically and biologically speaking. Take them away, and you face a sterile existence with no highs or lows to speak of. Emotionless, you will simply exist, rather than live. Tragically, this is the very kind of reality many narcoleptic patients are forced to adopt for reasons we will now explore.

Medically, narcolepsy is considered to be a neurological disorder, meaning that its origins are within the central nervous system, specifically the brain. The condition usually emerges between ages ten and twenty years. There is some genetic basis to narcolepsy, but it is not inherited. Instead, the genetic cause appears to be a mutation, so the disorder is not passed from parent to child. However, gene mutations, at least as we currently understand them in the context of this disorder, do not explain all incidences of narcolepsy. Other triggers remain to be identified. Narcolepsy is also not unique to humans, with numerous other mammals expressing the disorder.

There are at least three core symptoms that make up the disorder: (1) excessive daytime sleepiness, (2) sleep paralysis, and (3) cataplexy.

The first symptom of excessive daytime sleepiness is often the most disruptive and problematic to the quality of day-to-day life for narcoleptic patients. It involves daytime sleep attacks: overwhelming, utterly irresistible urges to sleep at times when you want to be awake, such as working at your desk, driving, or eating a meal with family or friends.

Having read that sentence, I suspect many of you are thinking, "Oh my goodness, I have narcolepsy!" That is unlikely. It is far more probable that you are suffering from chronic sleep deprivation. About one in every 2,000 people suffers from narcolepsy, making it about as common as multiple sclerosis. The sleep attacks that typify excessive daytime sleepiness are usually the first symptom to appear. Just to give you a sense of what that feeling is, relative to what you may be considering, it would be the sleepiness equivalent of staying awake for three to four days straight.

The second symptom of narcolepsy is sleep paralysis: the frightening loss of ability to talk or move when waking up from sleep. In essence, you become temporarily locked in your body.

Most of these events occur in REM sleep. You will remember that during REM sleep, the brain paralyzes the body to keep you from acting out your dreams. Normally, when we wake out of a dream, the brain releases the body from the paralysis in perfect synchrony, right at the moment when waking consciousness returns. However, there can be rare occasions when the paralysis of the REM state lingers on despite the brain having terminated sleep, rather like that last guest at a party who seems unwilling to recognize the event is over and it is time to leave the premises. As a result, you begin to wake up, but you are unable to lift your eyelids, turn over, cry out, or move any of the muscles that control your limbs. Gradually, the paralysis of REM sleep does wear off, and you regain control of your body, including your eyelids, arms, legs, and mouth.

Don't worry if you have had an episode of sleep paralysis at some point in your life. It is not unique to narcolepsy. Around one in four healthy individuals will experience sleep paralysis, which is to say that it is as common as hiccups. I myself have experienced sleep paralysis several times, and I do not suffer from narcolepsy. Narcoleptic patients will, however, experience sleep paralysis far more frequently and severely than healthy individuals. This nevertheless means that sleep paralysis is a symptom associated with narcolepsy, but it is not unique to narcolepsy.

A brief detour of an otherworldly kind is in order at this moment. When individuals undergo a sleep paralysis episode, it is often associated with feelings of dread and a sense of an intruder being present in the room. The fear comes from an inability to act in response to the perceived threat, such as not being able to shout out, stand up and leave the room, or prepare to defend oneself. It is this set of features of sleep paralysis that we now believe explains a large majority of alien abduction claims. Rarely do you hear of aliens accosting an individual in the middle of the day with testimonial witnesses standing in plain sight, dumbstruck by the extraterrestrial kidnapping in progress. Instead, most alleged alien abductions take place at night; most classic alien visitations in Hollywood movies like *Close Encounters of the Third Kind* or *E.T.* also occur at night. Moreover, victims of claimed alien abductions frequently report the sense of, or real presence of, a being in the room (the alien). Finally—and this is the key giveaway—the alleged victim frequently describes having been injected with a "paralyzing agent." Consequently, the victim will describe wanting to fight back, run away, or call out for help but being unable to do so. The offending force is, of course, not aliens, but the persistence of REM-sleep paralysis upon awakening.

The third and most astonishing core symptom of narcolepsy is called cataplexy. The word comes from the Greek *kata*, meaning down, and *plexis*, meaning a stroke or seizure—that is, a falling-down seizure. However, a cataplectic attack is not a seizure at all, but rather a sudden loss of muscle control. This can range from slight weakness wherein the head droops, the face sags, the jaw drops, and speech becomes slurred to a buckling of knees or a sudden and immediate loss of all muscle tone, resulting in total collapse on the spot.

You may be old enough to remember a child's toy that involved an animal, often a donkey, standing on a small, palm-sized pedestal with a button underneath. It was similar to a puppet on strings, except that the strings were not attached to the outside limbs, but rather woven through the limbs on the inside, and connected to the button underneath. Depressing the button relaxed the inner string tension, and the donkey would collapse into a heap. Release the button, pulling the inner strings taut, and the donkey would snap back upright to firm attention. The demolition of muscle tone that occurs during a full-blown cataplectic attack, resulting in total body collapse, is very much like this toy, but the consequences are no laughing matter.

If this were not wicked enough, there is an extra layer of malevolence to the condition that truly devastates the patient's quality of life. Cataplectic attacks are not random, but are triggered by moderate or strong emotions, positive or negative. Tell a funny joke to a narcoleptic patient, and they may literally collapse in front of you. Walk into a room and surprise a patient, perhaps while they are chopping food with a sharp knife, and they will collapse perilously. Even standing in a nice warm shower can be enough of a pleasurable experience to cause a patient's legs to buckle and have a potentially dangerous fall caused by the cataplectic muscle loss.

Now extrapolate this, and consider the dangers of driving a car and being startled by a loud horn. Or playing an enjoyable game with your children, or having them jump on you and tickle you, or feeling strong, tear-welling joy at one of their musical school recitals. In a narcoleptic patient with cataplexy, any one of these may cause the sufferer to collapse into the immobilized prison of his or her own body. Consider, then, how difficult it is to have a loving, pleasurable sexual relationship with a narcoleptic partner. The list becomes endless, with predictable and heart-wrenching outcomes.

Unless patients are willing to accept these crumpling attacks, which is really no option of any kind, all hope of living an emotionally fulfilling life must be abandoned. A narcoleptic patient is banished to a monotonic existence of emotional neutrality. They must forfeit any semblance of succulent emotions that we are all nourished by on a moment-to-moment basis. It is the dietary equivalent of eating the same tepid bowl of unflavorful porridge day after day. You can well imagine the loss of appetite for such a life.

If you saw a patient collapse under the influence of cataplexy, you would be convinced that they had fallen completely unconscious or into a powerful sleep. This is untrue. Patients are awake and continue to perceive the outside world around them. Instead, what the strong emotion has triggered is the total (or sometimes partial) body paralysis of REM sleep without the sleep of the REM state itself. Cataplexy is therefore an abnormal functioning of the REM-sleep circuitry within the brain, wherein one of its features—muscle atonia—is inappropriately deployed while the individual is awake and behaving, rather than asleep and dreaming.

We can of course explain this to an adult patient, lowering their anxiety during the event through comprehension of what is happening, and help them rein in or avoid emotional highs and lows to reduce cataplectic occurrences. However, this is much more difficult in a ten-year-old youngster. How can you explain such a villainous symptom and disorder to a child with narcolepsy? And how do you prevent a child from enjoying the normal roller coaster of emotional existence that is a natural and integral part of a growing life and developing brain? Which is to say, how do you prevent a child from being a child? There are no easy answers to these questions.

We are, however, beginning to discover the neurological basis of narcolepsy and, in conjunction, more about healthy sleep itself. In chapter 3, I described the parts of the brain involved in the maintenance of normal wakefulness: the alerting, activating regions of the brain stem and the sensory gate of the thalamus that sits on top, a setup that looks almost like a scoop of ice cream (thalamus) on a cone (brain stem). As the brain stem powers down at night, it removes its stimulating influence to the sensory gate of the thalamus. With the closing of the sensory gate, we stop perceiving the outside world, and thus we fall asleep.

What I did not tell you, however, was how the brain stem knows that it's time to turn off the lights, so to speak, and power down wakefulness to begin sleep. Something has to switch the activating influence of the brain stem off, and in doing so, allow sleep to be switched on. That switch—the sleep-wake switch—is located just below the thalamus in the center of the brain, in a region called the hypothalamus. It is the same neighborhood that houses the twenty-four-hour master biological clock, perhaps unsurprisingly.

The sleep-wake switch within the hypothalamus has a direct line of communication to the power station regions of the brain stem. Like an electrical light switch, it can flip the power on (wake) or off (sleep). To do this, the sleep-wake switch in the hypothalamus releases a neurotransmitter called orexin. You can think of orexin as the chemical finger that flips the switch to the "on," wakefulness, position. When orexin is released down onto your brain stem, the switch has been unambiguously flipped, powering up the wakefulness-generating centers of the brain stem. Once activated by the switch, the brain stem pushes open the sensory gate of the thalamus, allowing the perceptual world to flood into your brain, transitioning you to full, stable wakefulness.

At night, the opposite happens. The sleep-wake switch stops releasing orexin onto the brain stem. The chemical finger has now flipped the switch to the "off" position, shutting down the rousing influence from the power station of the brain stem. The sensory business being conducted within the thalamus is closed down by a sealing of the sensory gate. We lose perceptual contact with the outside world, and now sleep. Lights off, lights on, lights off, lights on—this is the neurobiological job of the sleep-wake switch in the hypothalamus, controlled by orexin.

Ask an engineer what the essential properties of a basic electrical switch are, and they will inform you of an imperative: the switch must be definitive. It must either be fully on or fully off—a binary state. It must not float in a wishy-washy manner between the "on" and "off" positions. Otherwise, the electrical system will not be stable or predictable. Unfortunately, this is exactly what happens to the sleep-wake switch in the disorder of narcolepsy, caused by marked abnormalities of orexin.

Scientists have examined the brains of narcoleptic patients in painstaking detail after they have passed away. During these postmortem investigations, they discovered a loss of almost 90 percent of all the cells that produce orexin. Worse still, the welcome sites, or receptors, of orexin that cover the surface of the power station of the brain stem were significantly reduced in number in narcoleptic patients, relative to normal individuals.

Because of this lack of orexin, made worse by the reduced number of receptor sites to receive what little orexin does drip down, the sleep-wake state of the narcoleptic brain is unstable, like a faulty flip-flop switch. Never definitively on or off, the brain of a narcoleptic patient wobbles precariously around a middle point, teeter-tottering between sleep and wakefulness.

The orexin-deficient state of this sleep-wake system is the main cause of the first and primary symptom of narcolepsy, which is excessive daytime sleepiness and the surprise attacks of sleep that can happen at any moment. Without the strong finger of orexin pushing the sleep-wake switch all the way over into a definitive "on" position, narcoleptic patients cannot sustain resolute wakefulness throughout the day. For the same reasons, narcoleptic patients have terrible sleep at night, dipping into and out of slumber in choppy fashion. Like a faulty light switch that endlessly flickers on and off, day and night, so goes the erratic sleep and wake experience suffered by a narcoleptic patient across each and every twenty-four-hour period.

Despite wonderful work by many of my colleagues, narcolepsy currently represents a failure of sleep research at the level of effective treatments. While we have effective interventions for other sleep disorders, such as insomnia and sleep apnea, we lag far behind the curve for treating narcolepsy. This is in part due to the rarity of the condition, making it unprofitable for drug companies to invest their research effort, which is often a driver of fast treatment progress in medicine.

For the first symptom of narcolepsy—daytime sleep attacks—the only treatment used to be high doses of the wake-promoting drug amphetamine. But amphetamine is powerfully addictive. It is also a "dirty" drug, meaning that it is promiscuous and affects many different chemical systems in the brain and body, leading to terrible side effects. A newer, "cleaner" drug, called Provigil, is now used to help narcoleptic patients stay more stably awake during the day and has fewer downsides. Yet it is marginally effective.

Antidepressants are often prescribed to help with the second and third symptoms of narcolepsy—sleep paralysis and cataplexy—as they suppress REM sleep, and it is REM-sleep paralysis that is integral to these two symptoms. Nevertheless, antidepressants simply lower the incidence of both; they do not eradicate them.

Overall, the treatment outlook for narcoleptic patients is bleak at present, and there is no cure in sight. Much of the treatment fate of narcolepsy sufferers and their families resides in the slower-progressing hands of academic research, rather than the more rapid progression of big pharmaceutical companies. For now, patients simply must try to manage life with the disorder, living as best they can.

Some of you may have had the same realization that several drug companies did when we learned about the role of orexin and the sleep-wake switch in narcolepsy: could we reverse-engineer the knowledge and, rather than enhance orexin to give narcoleptic patients more stable wakefulness during the day, try and shut it off at night, thereby offering a novel way of inducing sleep in insomnia patients? Pharmaceutical companies are indeed trying to develop compounds that can block orexin at night, forcing it to flip the switch to the "off" position, potentially inducing more naturalistic sleep than the problematic and sedating sleep drugs we currently have.

Unfortunately, the first of these drugs, suvorexant (brand name Belsomra), has not proved to be the magic bullet many hoped. Patients in the FDA-mandated clinical trials fell asleep just six minutes faster than those taking a placebo. While future formulations may prove more efficacious, non-pharmacological methods for the treatment of insomnia, outlined in the next chapter, remain a far superior option for insomnia sufferers.

FATAL FAMILIAL INSOMNIA

Michael Corke became the man who could not sleep—and the condition took his life. Before the insomnia took hold, Corke was a high-functioning, active individual, a devoted husband, and a teacher of music at a high school in New Lexon, just south of Chicago. At age forty he began having trouble sleeping. At first, Corke felt that his wife's snoring was to blame. In response to this suggestion, Penny Corke decided to sleep on the couch for the next ten nights. Corke's insomnia did not abate, and only became worse. After months of poor sleep, and realizing the cause lay elsewhere, Corke decided to seek medical help. None of the doctors who first examined Corke could identify the trigger of his insomnia, and some diagnosed him with sleep-unrelated disorders, such as multiple sclerosis.

Corke's insomnia eventually progressed to the point where he was completely unable to sleep. Not a wink. No mild sleep medications or even heavy sedatives could wrestle his brain from the grip of permanent wakefulness. Should you have observed Corke at this time, it would be clear how desperate he was for sleep. His eyes would make your own feel tired. His blinks were achingly slow, as if the eyelids wanted to stay shut, mid-blink, and not reopen for days. They telegraphed the most despairing hunger for sleep you could imagine.

After eight straight weeks of no sleep, Corke's mental faculties were quickly fading. This cognitive decline was matched in speed by the rapid deterioration of his body. So compromised were his motor skills that even coordinated walking became difficult. One evening Corke was to conduct a school orchestral performance. It took several painful (though heroic) minutes for him to complete the short walk through the orchestra and climb atop the conductor's rostrum, all cane-assisted.

As Corke approached the six-month mark of no sleep, he was bedridden and approaching death. Despite his young age, Corke's neurological condition resembled that of an elderly individual in the end stages of dementia. He could not bathe or clothe himself. Hallucinations and delusions were rife. His ability to generate language was all but gone, and he was resigned to communicating through rudimentary head movements and rare inarticulate utterances whenever he could muster the energy. Several more months of no sleep and Corke's body and mental faculties shut down completely. Soon after turning forty-two years old, Michael Corke died of a rare, genetically inherited disorder called fatal familial insomnia (FFI). There are no cures for this disorder. It is one of the most mysterious conditions in the annals of medicine.

The underlying cause of FFI is increasingly well understood, and builds on much of what we have discussed regarding the normal mechanisms of sleep generation. The culprit is an anomaly of a gene called PrNP, which stands for prion protein. All of us have prion proteins in our brain, and they perform useful functions. However, a rogue version of the protein is triggered by this genetic defect, resulting in a mutated version that spreads like a virus.* In this genetically crooked form, the protein begins targeting and destroying certain parts of the brain, resulting in a rapidly accelerating form of brain degeneration as the protein spreads.

One region that this malfeasant protein attacks, and attacks comprehensively, is the thalamus—that sensory gate within the brain that must close shut for wakefulness to end and sleep to begin. When scientists performed postmortem examinations of the brains of early sufferers of FFI, they discovered a thalamus that was peppered with holes, almost like a block of Swiss cheese. The prion proteins had burrowed throughout the thalamus, utterly degrading its structural integrity. This

*Fatal familial insomnia is part of a family of prion protein disorders that also includes Creutzfeldt-Jakob disease, or so-called mad cow disease, though the latter involves the destruction of different regions of the brain not strongly associated with sleep.

was especially true of the outer layers of the thalamus, which form the sensory doors that should close shut each night.

Due to this puncturing attack by the prion proteins, the sensory gate of the thalamus was effectively stuck in a permanent "open" position. Patients may not be able to switch off their conscious perception of the outside world and, as a result, cannot drift off into the sleep that they so desperately needed. In addition, the signals sent from the brain down into the body that prepare us for sleep—the reduction of heart rate, blood pressure, and metabolism, and the lowering of core body temperature—all must pass through the thalamus on their way down the spinal cord, and are then mailed out to the different tissues and organs of the body. But those signals were thwarted by the damage to the thalamus, adding to the impossibility of sleep in the patients.

Current treatment prospects are few. There has been some interest in an antibiotic called doxycycline, which seems to slow the rate of the rogue protein accumulation in other prion disorders, such as Creutzfeldt-Jakob disease, or so-called mad cow disease. Clinical trials for this potential therapy are now getting under way.

Beyond the race for a treatment and cure, an ethical issue emerges in the context of the disease. Since FFI is genetically inherited, we have been able to retrospectively trace some of its legacy through generations. That genetic lineage runs all the way back into Europe, and specifically Italy, where a number of afflicted families live. Careful detective work has rolled the genetic timeline back further, to a Venetian doctor in the late eighteenth century who appeared to have a clear case of the disorder. Undoubtedly, the gene goes back even further than this individual. More important than tracing the disease's past, however, is predicting its future. The genetic certainty raises a eugenically fraught question: If your family's genes mean that you could one day be struck down by the fatal inability to sleep, would you want to be told your fate? Furthermore, if you know that fate and have not yet had children, would that change your decision to do so, knowing you are a gene carrier and that you have the potential to prevent a next-step transmission of the disease? There are no simple answers, certainly none that science can (or perhaps should) offer—an additionally cruel tendril of an already heinous condition.

SLEEP DEPRIVATION VS. FOOD DEPRIVATION

FFI suggests that a lack of sleep could possibly kill a human being. Scientifically, however, it remains arguably inconclusive, as there may be other disease-related processes that could contribute to death, and they are hard to distinguish from those of a lack of sleep. There have been individual case reports of humans dying as a result of prolonged total sleep deprivation, such as Jiang Xiaoshan. He was alleged to have stayed awake for eleven days straight to watch all the games of the 2012 European soccer championships, all the while working at his job each day. On day 12, Xiaoshan was found dead in his apartment by his mother from an apparent lack of sleep. Then there was the tragic death of a Bank of America intern, Moritz Erhardt, who suffered a life-ending epileptic seizure after acute sleep deprivation from the work overload that is so endemic and expected in that profession, especially from the juniors in such organizations. Nevertheless, these are simply case studies, and they are hard to validate and scientifically verify after the fact.

Research studies in animals have, however, provided definitive evidence of the deadly nature of total sleep deprivation, free of any comorbid disease. The most dramatic, disturbing, and ethically provoking of these studies was published in 1983 by a research team at the University of Chicago. Their experimental question was simple: Is sleep necessary for life? By preventing rats from sleeping for weeks on end in a gruesome ordeal, they came up with an unequivocal answer: rats will die after eighteen days without sleep, on average.

Two additional results quickly followed. First, death ensued as quickly from total sleep deprivation as it did from total food deprivation. Second, rats selectively deprived of REM sleep died within twenty-seven to fifty-four days. A total absence of NREM sleep still proved fatal, it just took longer to inflict the same mortal consequence—within twenty-eight to sixty-six days.

There was, however, an issue. Unlike starvation, where the cause of death is easily identified, the researchers could not determine why the rats had died following sleep's absence, despite how quickly death had arrived. Some hints emerged from assessments made during the experiment, as well as the later postmortems.

First, despite eating far more than their sleep-rested counterparts, the sleep-deprived rats rapidly began losing body mass during the study. Second, they could no longer regulate their core body temperature. The more sleep-deprived the rats were, the colder they became, regressing toward ambient room temperature. This was a perilous state to be in. All mammals, humans included, live on the edge of a thermal cliff. Physiological processes within the mammalian body can only operate within a remarkably narrow temperature range. Dropping below or above these life-defining thermal thresholds is a fast track to death.

It was no coincidence that these metabolic and thermal consequences were jointly occurring. When core body temperature drops, mammals respond by increasing their metabolic rate. Burning energy releases heat to warm the brain and body to get them back above the critical thermal threshold so as to avert death. But it was a futile effort in the rats lacking sleep. Like an old wood-burning stove whose top vent has been left open, no matter how much fuel was being added to the fire, the heat simply flew out the top. The rats were effectively metabolizing themselves from the inside out in response to hypothermia.

The third, and perhaps most telling, consequence of sleep loss was skin deep. The privation of sleep had left these rats literally threadbare. Sores had appeared across the rats' skin, together with wounds on their paws and tails. Not only was the metabolic system of the rats starting to implode, but so, too, was their immune system.* They could not fend off even the most basic of infections at their epidermis—or below it, as we shall see.

If these outward signs of degrading health were not shocking

*The senior scientist conducting these studies, Allan Rechtschaffen, was once contacted by a well-known women's fashion magazine after these findings were published. The writer of the article wanted to know if total sleep deprivation offered an exciting, new, and effective way for women to lose weight. Struggling to comprehend the audacity of what had been asked of him, Rechtschaffen attempted to compose a response. Apparently, he admitted that enforced total sleep deprivation in rats results in weight loss, so yes, acute sleep deprivation for days on end does lead to weight loss. The writer was thrilled to get the story line they wanted. However, Rechtschaffen offered a footnote: that in combination with the remarkable weight loss came skin wounds that wept lymph fluid, sores that had eviscerated the rats' feet, a decrepitude that resembled accelerated aging, together with catastrophic (and ultimately fatal) internal organ and immune-system collapse "just in case appearance, and a longer life, were also part of your readers' goals." Apparently, the interview was terminated soon after.

enough, the internal damage revealed by the final postmortem was equally ghastly. A landscape of utter physiological distress awaited the pathologist. Complications ranged from fluid in the lungs and internal hemorrhaging to ulcers puncturing the stomach lining. Some organs, such as the liver, spleen, and kidneys, had physically decreased in size and weight. Others, like the adrenal glands that respond to infection and stress, were markedly enlarged. Circulating levels of the anxiety-related hormone corticosterone, released by the adrenal glands, had spiked in the sleepless rats.

What, then, was the cause of death? Therein lay the issue: the scientists had no idea. Not all the rats suffered the same pathological signature of demise. The only commonality across the rats was death itself (or the high likelihood of it, at which point the researchers euthanized the animals).

In the years that followed, further experiments—the last of their kind, as scientists felt (rightly, in my personal view) uneasy about the ethics of such experiments based on the outcome—finally resolved the mystery. The fatal final straw turned out to be septicemia—a toxic and systemic (whole organism) bacterial infection that coursed through the rats' bloodstream and ravaged the entire body until death. Far from a vicious infection that came from the outside, however, it was simple bacteria from the rats' very own gut that inflicted the mortal blow—one that an otherwise healthy immune system would have easily quelled when fortified by sleep.

The Russian scientist Marie de Manacéïne had in fact reported the same mortal consequences of continuous sleep deprivation in the medical literature a century earlier. She noted that young dogs died within several days if prevented from sleeping (which are difficult studies for me to read, I must confess). Several years after de Manacéïne's studies, Italian researchers described equally lethal effects of total sleep deprivation in dogs, adding the observation of neural degeneration in the brain and spinal cord at postmortem.

It took another hundred years after the experiments of de Manacéïne, and the advancements in precise experimental laboratory assessments, before the scientists at the University of Chicago finally uncovered why life ends so quickly in the absence of sleep. Perhaps you have seen that small plastic red box on the walls of extremely hazardous work environments that has the following words written on the front: "Break glass in case of emergency." If you impose a total absence of sleep on an organism, rat or human, it indeed becomes an emergency, and you will find the biological equivalent of this shattered glass strewn throughout the brain and the body, to fatal effect. This we finally understand.

NO, WAIT—YOU ONLY NEED 6.75 HOURS OF SLEEP!

Reflecting on these deathly consequences of long-term/chronic and short-term/acute sleep deprivation allows us to address a recent controversy in the field of sleep research—one that many a newspaper, not to mention some scientists, apprehended incorrectly. The study in question was conducted by researchers at the University of California, Los Angeles, on the sleep habits of specific pre-industrial tribes. Using wristwatch activity devices, the researchers tracked the sleep of three hunter-gatherer tribes that are largely untouched by the ways of industrial modernity: the Tsimané people in South America, and the San and Hadza tribes in Africa, which we have previously discussed. Assessing sleep and wake times day after day across many months, the findings were thus: tribespeople averaged just 6 hours of sleep in the summer, and about 7.2 hours of sleep in the winter.

Well-respected media outlets touted the findings as proof that human beings do not, after all, need a full eight hours of sleep, some suggesting we can survive just fine on six hours or less. For example, the headline of one prominent US newspaper read:

"Sleep Study on Modern-Day Hunter-Gatherers Dispels Notion That We're Wired to Need 8 Hours a Day."

Others started out with the already incorrect assumption that modern societies need only seven hours of sleep, and then questioned whether we even need that much: "Do We Really Need to Sleep 7 Hours a Night?"

How can such prestigious and well-respected entities reach these conclusions, especially after the science that I have presented in this chapter? Let us carefully reevaluate the findings, and see if we still arrive at the same conclusion.

First, when you read the paper, you will learn that the tribespeople were actually giving themselves a 7- to 8.5-hour sleep opportunity each night. Moreover, the wristwatch device, which is neither a precise nor gold standard measure of sleep, estimated a range of 6 to 7.5 hours of this time was spent asleep. The sleep opportunity that these tribespeople provide themselves is therefore almost identical to what the National Sleep Foundation and the Centers for Disease Control and Prevention recommend for all adult humans: 7 to 9 hours of time in bed.

The problem is that some people confuse time slept with sleep opportunity time. We know that many individuals in the modern world only give themselves 5 to 6.5 hours of sleep opportunity, which normally means they will only obtain around 4.5 to 6 hours of actual sleep. So no, the finding does not prove that the sleep of hunter-gatherer tribes is similar to ours in the post-industrial era. They, unlike us, give themselves more sleep opportunity than we do.

Second, let us assume that the wristwatch measurements are perfectly accurate, and that these tribes obtain an annual average of just 6.75 hours of sleep. The next erroneous conclusion drawn from the findings was that humans must, therefore, naturally need a mere 6.75 hours of sleep, and no more. Therein lies the rub.

If you refer back to the two newspaper headlines I quoted, you'll notice they both use the word "need." But what *need* are we talking about? The (incorrect) presupposition made was this: whatever sleep the tribespeople were obtaining is all that a human *needs*. It is flawed reasoning on two counts. *Need* is not defined by that which is obtained (as the disorder of insomnia teaches us), but rather whether or not that amount of sleep is sufficient to accomplish all that sleep does. The most obvious *need*, then, would be for life—and healthy life. Now we discover that the average life span of these hunter-gatherers is just fifty-eight years, even though they are far more physically active than we are, rarely obese, and are not plagued by the assault of processed foods that erode our health. Of course, they do not have access to modern medicine and sanitation, both of which are reasons that many of us in industrialized, first-world nations have an expected life span that exceeds theirs by over a decade. But it is telling that, based on epidemiological data, any adult sleeping an average of 6.75 hours a night would be predicted to live only into their early sixties: very close to the median life span of these tribespeople.

More prescient, however, is what normally kills people in these tribes. So long as they survive high rates of infant mortality and make it through adolescence, a common cause of death in adulthood is infection. Weak immune systems are a known consequence of insufficient sleep, as we have discussed in great detail. I should also note that one of the most common immune system failures that kills individuals in hunter-gatherer clans are intestinal infections—something that shares an intriguing overlap with the deadly intestinal tract infections that killed the sleep-deprived rats in the above studies.

Recognizing this shorter life span, which fits well with the acclaimed shorter sleep amounts the researchers measured, the next error in logic many made is exposed by asking *why* these tribes would sleep what appears to be too little, based on all that we know from thousands of research studies.

We do not yet know of all the reasons, but a likely contributing factor lies in the title we apply to these tribes: hunter-gatherers. One of the few universal ways of forcing animals of all kinds to sleep less than normal amounts is to limit food, applying a degree of starvation. When food becomes scarce, sleep becomes scarce, as animals try to stay awake longer to forage. Part of the reason that these hunter-gatherer tribes are not obese is because they are constantly searching for food, which is never abundant for long stretches. They spend much of their waking lives in pursuit and preparation of nutrition. For example, the Hadza will face days where they obtain 1,400 calories or less, and routinely eat 300 to 600 fewer daily calories than those of us in modern Western cultures. A large proportion of their year is therefore spent in a state of lower-level starvation, one that can trigger well-characterized biological pathways that reduce sleep time, even though sleep *need* remains higher than that obtained if food were abundant. Concluding that humans, modern-living or pre-industrial, *need* less than seven hours of sleep therefore appears to be an erroneous conclusion.

IS SLEEPING NINE HOURS A NIGHT TOO MUCH?

Epidemiological evidence suggests that the relationship between sleep and mortality risk is not linear, such that the more and more sleep you get, the lower and lower your death risk (and vice versa). Rather, there is an upward hook in death risk once the average sleep amount passes nine hours.

Two points are worthy of mention in this regard. First, should you explore those studies in detail, you learn that the causes of death in individuals sleeping nine hours or longer include infection (e.g., pneumonia) and immune-activating cancers. We know from evidence discussed earlier in the book that sickness, especially sickness that activates a powerful immune response, activates more sleep. Ergo, the sickest individuals should be sleeping longer to battle back against illness using the suite of health tools sleep has on offer. It is simply that some illnesses, such as cancer, can be too powerful even for the mighty force of sleep to overcome, no matter how much sleep is obtained. The illusion created is that too much sleep leads to an early death, rather than the sickness being too much despite all efforts to the contrary from the beneficial sleep extension.

Second, leading sleep experts have suggested that the mortality risk associated with long sleep (the far-right side of the U curve) is, in part, a reflection of a poor-*quality* sleep, rather than *too much* sleep. Poor-quality sleep is significantly linked with a higher mortality risk, and poor-quality sleep can cause people to stay in bed for longer. Thus, poor quality of sleep, which can masquerade as long sleep, and not long sleep itself, helps explain that far-right side of the sleep and mortality curve.

Third, it is important not to overextend my point. I am not suggesting that sleeping eighteen or twenty-two hours each and every day, should that be physiologically possible, is more optimal than sleeping nine hours a day. Sleep is unlikely to operate in such a linear manner. Keep in mind that food, oxygen, and water are no different, and they, too, have a U-shaped relationship with mortality risk. Eating to excess shortens life. Extreme hydration can lead to fatal increases in blood pressure associated with stroke or heart attack. Too much oxygen in the blood, known as hyperoxia, is toxic to cells, especially those of the brain.

Sleep, like food, water, and oxygen, may share this relationship with mortality risk when taken to extremes. After all, wakefulness in the correct amount is evolutionarily adaptive, as is sleep. Both sleep and wake provide synergistic and critical, though often different, survival advantages. There is an adaptive balance to be struck between wakefulness and sleep. In humans, that appears to be around sixteen hours of total wakefulness, and between seven and nine hours of total sleep, for an average adult.

CHAPTER 13

iPads, Factory Whistles, and Nightcaps

What's Stopping You from Sleeping?

Many of us are beyond tired. Why? What, precisely, about modernity has so perverted our otherwise instinctual sleep patterns, eroded our freedom to sleep, and thwarted our ability to do so soundly across the night? For those of us who do not have a sleep disorder, the reasons underlying this state of sleep deficiency can seem hard to pinpoint—or, if seemingly clear, are erroneous.

Beyond longer commute times and "sleep procrastination" caused by late-evening television and digital entertainment—both of which are not unimportant in their top-and-tail snipping of our sleep time and that of our children—five key factors have changed how much and how well we sleep: (1) constant electric light as well as LED light, (2) regularized temperature, (3) caffeine (discussed in chapter 2), (4) alcohol, and (5) a legacy of punching time cards. It is this set of societally engineered forces that are responsible for many an individual's mistaken belief that they are suffering from medical insomnia.

THE DARK SIDE OF MODERN LIGHT

At 255–257 Pearl Street, in Lower Manhattan, not far from the Brooklyn Bridge, is the site of arguably the most unassuming yet seismic shift in our human history. Here Thomas Edison built the first power-generating station to support an electrified society. For the first time, the human race had a truly scalable method of unbuckling itself from our planet's natural twenty-four-hour cycle of light and dark. With a proverbial flick of a switch came a whimsical ability to control our environmental light and, with it, our wake and sleep phases. We, and not the rotating mechanics of planet Earth, would now decide when it was "night" and when it was "day." We are the only species that has managed to light the night to such dramatic effect.

Humans are predominantly visual creatures. More than a third of our brain is devoted to processing visual information, far exceeding that given over to sounds or smells, or those supporting language and movement. For early Homo sapiens, most of our activities would have ceased after the sun set. They had to, as they were predicated on vision, supported by daylight. The advent of fire, and its limited halo of light, offered an extension to post-dusk activities. But the effect was modest. In the early-evening glow of firelight, nominal social activities such as singing and storytelling have been documented in hunter-gatherer tribes like the Hadza and the San. Yet the practical limitations of firelight nullified any significant influence on the timing of our sleep-wake patterns.

Gas- and oil-burning lamps, and their forerunners, candles, offered a more forceful influence upon sustained nighttime activities. Gaze at a Renoir painting of nineteenth-century Parisian life and you will see the extended reach of artificial light. Spilling out of homes and onto the streets, gas lanterns began bathing entire city districts with illumination. In this moment, the influence of man-made light began its reengineering of human sleep patterns, and it would only escalate. The nocturnal rhythms of whole societies—not just individuals or single families—became quickly subject to light at night, and so began our advancing march toward later bedtimes.

For the suprachiasmatic nucleus—the master twenty-four-hour clock of the brain—the worst was yet to come. Edison's Manhattan power station enabled the mass adoption of incandescent light. Edison did not create the first incandescent lightbulb—that honor went to the English chemist Humphry Davy in 1802. But in the mid-1870s, Edison Electric Light Company began developing a reliable, mass-marketable lightbulb. Incandescent light bulbs, and decades later, fluorescent light bulbs, guaranteed that modern humans would no longer spend much of the night in darkness, as we had for millennia past.

One hundred years post-Edison, we now understand the biological mechanisms by which the electric lightbulbs managed to veto our natural timing and quality of sleep. The visible light spectrum—that which our eyes can see—runs the gamut from shorter wavelengths (approximately 380 nanometers) that we perceive as cooler violets and blues, to the longer wavelengths (around 700 nanometers) that we sense as warmer yellows and reds. Sunlight contains a powerful blend of all of these colors, and those in between (as the iconic Pink Floyd album cover of *Dark Side of the Moon* illuminates [so to speak]).

Before Edison, and before gas and oil lamps, the setting sun would take with it this full stream of daylight from our eyes, sensed by the twenty-four-hour clock within the brain (the suprachiasmatic nucleus, described in chapter 2). The loss of daylight informs our suprachiasmatic nucleus that nighttime is now in session; time to release the brake pedal on our pineal gland, allowing it to unleash vast quantities of melatonin that signal to our brains and bodies that darkness has arrived and it is time for bed. Appropriately scheduled tiredness, followed by sleep, would normally occur several hours after dusk across our human collective.

Electric light put an end to this natural order of things. It redefined the meaning of midnight for generations thereafter. Artificial evening light, even that of modest strength, or lux, will fool your suprachiasmatic nucleus into believing the sun has not yet set. The brake on melatonin, which should otherwise have been released with the timing of dusk, remains forcefully applied within your brain under duress of electric light.

The artificial light that bathes our modern indoor worlds will therefore halt the forward progress of biological time that is normally signaled by the evening surge in melatonin. Sleep in modern humans is delayed from taking off the evening runway, which would naturally occur somewhere between eight and ten p.m., just as we observe in hunter-gatherer tribes. Artificial light in modern societies thus tricks us into believing night is still day, and does so using a physiological lie.

The degree to which evening electric light winds back your internal twenty-four-hour clock is important: usually two to three hours each evening, on average. To contextualize that, let's say you are reading this book at eleven p.m. in New York City, having been surrounded by electric light all evening. Your bedside clock may be registering eleven p.m., but the omnipresence of artificial light has paused the internal tick-tocking of time by hindering the release of melatonin. Biologically speaking, you've been dragged westward across the continent to the internal equivalent of Chicago time (ten p.m.), or even San Francisco time (eight p.m.).

Artificial evening and nighttime light can therefore mimic sleep-onset insomnia—the inability to fall asleep within twenty-five minutes. By delaying the release of melatonin, artificial evening light makes it considerably less likely that you'll be able to fall asleep at a reasonable time. When you do finally turn out the bedside light, hoping that sleep will come quickly is made all the more difficult. It will be some time before the rising tide of melatonin is able to submerge your brain and body in peak concentrations, instructed by the darkness that only now has begun—in other words, before you are biologically capable of organizing the onset of robust, stable sleep.

What of a petite bedside lamp? How much can that really influence your suprachiasmatic nucleus? Some, it turns out. Even a hint of dim light—8 to 10 lux—has been shown to delay the release of nighttime melatonin in humans. A dim bedside lamp may emit twice as much: anywhere from 20 to 80 lux. A subtly lit living room, where most people reside in the hours before bed, will hum at around 200 lux. Despite being just 1 to 2 percent of the strength of daylight, this ambient level of incandescent home lighting can have a significant melatonin-suppressing influence within the brain.

Just when things looked as bad as they could get for the suprachiasmatic nucleus with incandescent lamps, a new invention in 1997 made the situation far worse: blue light–emitting diodes, or blue LEDs. For this invention, Shuji Nakamura, Isamu Akasaki, and Hiroshi Amano won the Nobel Prize in physics in 2014. It was a remarkable achievement. Blue LED lights offer considerable advantages over incandescent lamps in terms of lower energy demands and, for the lights themselves, longer life spans. But they may be inadvertently shortening and disrupting our own sleep.

The light receptors in the eye that communicate "daytime" to the suprachiasmatic nucleus are most sensitive to short-wavelength light within the blue spectrum—the exact sweet spot where blue LEDs are most powerful. As a consequence, evening blue LED light has a more harmful impact on human nighttime melatonin suppression than the warm, yellow light from old incandescent bulbs, even when their lux intensities are matched.

Of course, few of us stare headlong into the glare of an LED lamp each evening. But we do stare at LED-powered laptop screens, smartphones, and tablets each night, sometimes for many hours, often with these devices just feet or even inches away from our retinas. A recent survey of over fifteen hundred American adults found that 90 percent of individuals regularly used some form of portable electronic device sixty minutes or less before bedtime. It has a very real impact on your melatonin release, and thus ability to time the onset of sleep.

One of the earliest studies found that using an iPad—an electronic tablet enriched with blue LED light—for two hours prior to bed blocked the otherwise rising levels of melatonin by a significant 23 percent. A more recent report took the story several concerning steps further. Healthy adults lived for a two-week period in a tightly controlled laboratory environment. The two-week period was split in half, containing two different experimental arms that everyone passed through: (1) five nights of reading a book on an iPad for several hours before bed (no other iPad uses, such as email or Internet, were allowed), and (2) five nights of reading a printed paper book for several hours before bed, with the two conditions randomized in terms of which the participants experienced as first or second.

Compared to reading a printed book, reading on an iPad suppressed melatonin release by over 50 percent at night. Indeed, iPad reading delayed the rise of melatonin by up to three hours, relative to the natural rise in these same individuals when reading a printed book. When reading on the iPad, their melatonin peak, and thus instruction to sleep, did not occur until the early-morning hours, rather than before midnight. Unsurprisingly, individuals took longer to fall asleep after iPad reading relative to print-copy reading.

But did reading on the iPad actually change sleep quantity/quality above and beyond the timing of melatonin? It did, in three concerning ways. First, individuals lost significant amounts of REM sleep following iPad reading. Second, the research subjects felt less rested and sleepier throughout the day following iPad use at night. Third was a lingering aftereffect, with participants suffering a ninety-minute lag in their evening rising melatonin levels for several days after iPad use ceased—almost like a digital hangover effect.

Perhaps more than the blue light itself, research from Flinders University suggests that the harmful impact of evening smartphone and laptop use on sleep is due to a significant alerting effect. By artificially forcing us awake, thereby masking our natural tiredness at night, pre-bed technology keeps people awake for longer, and makes falling asleep more difficult. Using LED devices at night impacts our natural sleep rhythms, the quality of our sleep, and how alert we feel during the day. The societal and public health ramifications, discussed in the penultimate chapter, are not small. I, like many of you, have seen young children using electronic tablets at every opportunity throughout the day . . . and evening. The devices are a wonderful piece of technology. They enrich the lives and education of our youth. But such technology is also enriching their eyes and brains with powerful blue light and artificial stimulation that has a damaging effect on sleep—the sleep that young, developing brains so desperately need in order to flourish.*

Due to its omnipresence, solutions for limiting exposure to artificial evening light are challenging. A good start is to create lowered, dim light in the rooms where you spend your evening hours. Avoid powerful overhead lights. Mood lighting is the order of the night. Some committed individuals will even wear yellow-tinted glasses indoors in the afternoon and evening to help filter out the most harmful blue light that suppresses melatonin.

---

*For those wondering why cool blue light is the most potent of the visible light spectrum for regulating melatonin release, the answer lies in our distant ancestral past. Human beings, as we believe is true of all forms of terrestrial organisms, emerged from marine life. The ocean acts like a light filter, stripping away most of the longer, yellow and red wavelength light. What remains is the shorter, blue wavelength light. It is the reason the ocean, and our vision when submerged under its surface, appears blue. Much of marine life, therefore, evolved within the blue visible light spectrum, including the evolution of aquatic eyesight. Our biased sensitivity to cool blue light is a vestigial carryover from our marine forebears. Unfortunately, this evolutionary twist of fate has now come back to haunt us in a new era of blue LED light, discombobulating our melatonin rhythm and thus our sleep-wake rhythm.

Maintaining complete darkness throughout the night is equally critical, the easiest fix for which comes from blackout curtains. Finally, you can install software on your computers, phones, and tablet devices that gradually de-saturate the harmful blue LED light as evening progresses.

TURNING DOWN THE NIGHTCAP—ALCOHOL

Short of prescription sleeping pills, the most misunderstood of all "sleep aids" is alcohol. Many individuals believe alcohol helps them to fall asleep more easily, or even offers sounder sleep throughout the night. Both are resolutely untrue.

Alcohol is in a class of drugs called sedatives. It binds to receptors within the brain that prevent neurons from firing their electrical impulses. Saying that alcohol is a sedative often confuses people, as alcohol in moderate doses helps individuals liven up and become more social. How can a sedative enliven you? The answer comes down to the fact that your increased sociability is caused by sedation of one part of your brain, the prefrontal cortex, early in the timeline of alcohol's creeping effects. As we have discussed, this frontal lobe region of the human brain helps control our impulses and restrains our behavior. Alcohol immobilizes that part of our brain first. As a result, we "loosen up," becoming less controlled and more extroverted. But anatomically targeted brain sedation it still is.

Give alcohol a little more time, and it begins to sedate other parts of the brain, dragging them down into a stupefied state, just like the prefrontal cortex. You begin to feel sluggish as the inebriated torpor sets in. This is your brain slipping into sedation. Your desire and ability to remain conscious are decreasing, and you can let go of consciousness more easily. I am very deliberately avoiding the term "sleep," however, because sedation is not sleep. Alcohol sedates you out of wakefulness, but it does not induce natural sleep. The electrical brainwave state you enter via alcohol is not that of natural sleep; rather, it is akin to a light form of anesthesia.

Yet this is not the worst of it when considering the effects of the evening nightcap on your slumber. More than its artificial sedating influence, alcohol harms an individual's sleep in an additional two ways.

First, alcohol fragments sleep, littering the night with brief awakenings. Alcohol-infused sleep is therefore not continuous and, as a result, not restorative. Unfortunately, most of these nighttime awakenings go unnoticed by the sleeper since they don't remember them. Individuals therefore fail to link alcohol consumption the night before with feelings of next-day exhaustion caused by the undetected sleep disruption sandwiched in between. Keep an eye out for that coincidental relationship in yourself and/or others.

Second, alcohol will often suppress REM sleep, especially during the first half or two-thirds of the night. When the body metabolizes alcohol it produces by-product chemicals called aldehydes and ketones. The aldehydes in particular will block the brain's ability to generate REM sleep. It's rather like the cerebral version of cardiac arrest, preventing the pulsating beat of brainwaves that otherwise power dream sleep. People consuming even moderate amounts of alcohol in the afternoon and/or evening can inadvertently deprive themselves of dream sleep.

There is a sad and extreme demonstration of this fact observed in alcoholics who, when drinking, can show little in the way of any identifiable REM sleep. Going for such long stretches of time without dream sleep produces a tremendous buildup in, and backlog of, pressure to obtain REM sleep. So great, in fact, that it inflicts a frightening consequence upon these individuals: aggressive intrusions of dreaming while they are wide awake. The pent-up REM-sleep pressure erupts forcefully into waking consciousness, causing hallucinations, delusions, and gross disorientation. This process contributes to the psychotic state in alcoholics called "delirium tremens."*

Should the addict enter a rehabilitation program and abstain from alcohol, the brain will begin feasting on REM sleep, binging in a desperate effort to recover that which it has been long starved of—an effect called the REM-sleep rebound. We observe precisely the same consequences caused by excess REM-sleep pressure in individuals who have tried to break the sleep-deprivation world record (before this life-threatening feat was banned).

*V. Zarcone, "Alcoholism and sleep," *Advances in Bioscience and Biotechnology* 21 (1978): 29–38.

You don't have to be using alcohol to levels of abuse, however, to suffer its deleterious REM-sleep-disrupting consequences, as one study can attest. Recall that one function of REM sleep is to aid in memory integration and association: the type of information processing required for developing grammatical rules in new language learning, or in synthesizing large sets of related facts into an interconnected whole. To wit, researchers recruited a large group of college students for a seven-day study. The participants were assigned to one of three experimental conditions. On day 1, all the participants learned a novel, artificial grammar, rather like learning a new computer coding language or a new form of algebra. It was just the type of memory task that REM sleep is known to promote. Everyone learned the new material to a high degree of proficiency on that first day—around 90 percent accuracy. Then, a week later, the participants were tested to see how much of that information had been solidified by the six nights of intervening sleep.

What distinguished the three groups was the type of sleep they had. In the first group—the control condition—participants were allowed to sleep naturally and fully for all intervening nights. In the second group, the experimenters got the students a little drunk just before bed on the first night after daytime learning. They loaded up the participants with two to three shots of vodka mixed with orange juice, standardizing the specific blood alcohol amount on the basis of gender and body weight. In the third group, they allowed the participants to sleep naturally on the first and even the second night after learning, and then got them similarly drunk before bed on night 3.

Note that all three groups learned the material on day 1 while sober, and were tested while sober on day 7. This way, any difference in memory among the three groups could not be explained by the direct effects of alcohol on memory formation or later recall, but must be due to the disruption of memory that occurred in between.

On day 7, participants in the control condition remembered everything they had originally learned, even showing an enhancement of abstraction and retention of knowledge relative to initial levels of learning, just as we'd expect from good sleep. In contrast, those who had their sleep laced with alcohol on the first night after learning suffered what can conservatively be described as partial amnesia seven days later, forgetting more than 50 percent of all that original knowledge. This fits well with evidence we discussed earlier: that of the brain's non-negotiable requirement for sleep the first night after learning for the purposes of memory processing.

The real surprise came in the results of the third group of participants. Despite getting two full nights of natural sleep after initial learning, having their sleep doused with alcohol on the third night still resulted in almost the same degree of amnesia—40 percent of the knowledge they had worked so hard to establish on day 1 was forgotten.

The overnight work of REM sleep, which normally assimilates complex memory knowledge, had been interfered with by the alcohol. More surprising, perhaps, was the realization that the brain is not done processing that knowledge after the first night of sleep. Memories remain vulnerable to disruption of sleep (including that from alcohol) even up to three nights after learning, despite two full nights of natural sleep prior.

Framed practically, let's say that you are a student cramming for an exam on Monday. Diligently, you study all of the previous Wednesday. Your friends beckon you to come out that night for drinks, but you know how important sleep is, so you decline. On Thursday, friends again ask you to grab a few drinks in the evening, but to be safe, you turn them down and sleep soundly a second night. Finally, Friday rolls around—now three nights after your learning session—and everyone is heading out for a party and drinks. Surely, after being so dedicated to slumber across the first two nights after learning, you can now cut loose, knowing those memories have been safely secured and fully processed within your memory banks. Sadly, not so. Even now, alcohol consumption will wash away much of that which you learned and can abstract by blocking your REM sleep.

How long is it before those new memories are finally safe? We actually do not yet know, though we have studies under way that span many weeks. What we do know is that sleep has not finished tending to those newly planted memories by night 3. I elicit audible groans when I present these findings to my undergraduates in lectures. The politically incorrect advice I would (of course never) give is this: go to the pub for a drink in the morning. That way, the alcohol will be out of your system before sleep.

Glib advice aside, what is the recommendation when it comes to sleep and alcohol? It is hard not to sound puritanical, but the evidence is so strong regarding alcohol's harmful effects on sleep that to do otherwise would be doing you, and the science, a disservice. Many people enjoy a glass of wine with dinner. But it takes your liver and kidneys many hours to degrade and excrete that alcohol, even if you are an individual with fast-acting enzymes for ethanol decomposition. Nightly alcohol will likely disrupt your sleep, and the annoying advice of abstinence is the best, and most honest, I can offer.

GET THE NIGHTTIME CHILLS

Thermal environment, specifically the proximal temperature around your body and brain, is perhaps the most underappreciated factor determining the ease with which you will fall asleep tonight, and the quality of sleep you will obtain. Ambient room temperature, bedding, and nightclothes dictate the thermal envelope that wraps around your body at night. It is ambient room temperature that has suffered a dramatic assault from modernity. This change sharply differentiates the sleeping practices of modern humans from those of pre-industrial cultures, and from animals.

To successfully initiate and maintain sleep into the night, as described in chapter 2, your core temperature needs to decrease by 2 to 3 degrees Fahrenheit, or about 1 degree Celsius. For this reason, you will always find it easier to fall asleep in a room that is too cold than too hot, since a room that is too cold is at least dragging your brain and body in the correct (downward) temperature direction for sleep.

The decrease in core temperature is detected by a group of thermosensitive cells situated in the center of your brain within the hypothalamus. Those cells live right next door to the twenty-four-hour clock of the suprachiasmatic nucleus in the brain, and for good reason. Once core temperature dips below a threshold in the evening, the thermosensitive cells quickly deliver a neighborly message to the suprachiasmatic nucleus. The memo adds to that of naturally fading light, informing the suprachiasmatic nucleus to initiate the evening surge in melatonin, and with it, the timed ordering of sleep. Your nocturnal melatonin levels are therefore controlled not only by the loss of daylight at dusk, but also the drop in temperature that coincides with the setting sun. Environmental light and temperature therefore synergistically, though independently, dictate nightly melatonin levels and sculpt the ideal timing of sleep. In fact, if you bring an individual into the laboratory and allow them to freely choose when to go to bed, they will unwittingly pick the moment when their core body temperature is dropping most steeply.

Your body is not passive in letting the cool of night lull it into sleep, but actively participates. One way you control your core body temperature is using the surface of your skin. Most of the thermic work is performed by three parts of your body in particular: your hands, your feet, and your head. All three areas are rich in crisscrossing blood vessels, known as the arteriovenous anastomoses, that lie close to the skin's surface. Like stretching clothes over a drying line, this mass of vessels will allow blood to be spread across a large surface area of skin and come in close contact with the air that surrounds it. The hands, feet, and head are therefore remarkably efficient radiating devices that, just prior to sleep onset, jettison body heat in a massive thermal venting session so as to drop your core body temperature. Warm hands and feet help your body's core cool, inducing inviting sleep quickly and efficiently.

It is no evolutionary coincidence that we humans have developed the pre-bed ritual of splashing water on one of the most vascular parts of our bodies—our face, using one of the other highly vascular surfaces—our hands. You may think the feeling of being facially clean helps you sleep better, but facial cleanliness makes no difference to your slumber. The act itself does have sleep-inviting powers, however, as that water, warm or cold, helps dissipate heat from the surface of the skin as it evaporates, thereby cooling the inner body core.

The need to dump heat from our extremities is also the reason that you may occasionally stick your hands and feet out from underneath the bedcovers at night due to your core becoming too hot, usually without your knowing. Should you have children, you've probably seen the same phenomenon when you check in on them late at night: arms and legs dangling out of the bed in amusing (and endearing) ways, so different from the neatly positioned limbs you placed beneath the sheets upon first tucking them into bed. The limb rebellion aids in keeping the body core cool, allowing it to fall and stay asleep.

The coupled dependency between sleep and body cooling is evolutionarily linked to the twenty-four-hour ebb and flow of daily temperature. Homo sapiens (and thus modern sleep patterns) evolved in eastern equatorial regions of Africa. Despite experiencing only modest fluctuations in average temperature across a year (+/– 3°C, or 5.4°F), these areas have larger temperature differentials across a day and night in both the winter (+/– 14°F, or 8°C) and the summer (+/– 12°F, or 7°C).

Pre-industrial cultures, such as the nomadic Gabra tribe in northern Kenya, and the hunter-gatherers of the Hadza and San tribes, have remained in thermic harmony with this day-night cycle. They sleep in porous huts with no cooling or heating systems, minimal bedding, and lie semi-naked. They sleep this way from birth to death. Such willing exposure to ambient temperature fluctuations is a major factor (alongside the lack of artificial evening light) determining their well-timed, healthy sleep quality. Without indoor-temperature control, heavy bedding, or excess nighttime attire, they display a form of thermal liberalism that assists, rather than battles against, sleep's conditional needs.

In stark contrast, industrialized cultures have severed their relationship with this natural rise and fall of environmental temperature. Through climate-controlled homes with central heat and air-conditioning, and the use of bedcovers and pajamas, we have architected a minimally varying or even constant thermal tenor in our bedrooms. Bereft of the natural drop in evening temperature, our brains do not receive the cooling instruction within the hypothalamus that facilitates a naturally timed release of melatonin. Moreover, our skin has difficulty "breathing out" the heat it must in order to drop core temperature and make the transition to sleep, suffocated by the constant heat signal of controlled home temperatures.

A bedroom temperature of around 65 degrees Fahrenheit (18.3°C) is a reasonable goal for the sleep of most people, assuming standard bedding and clothing. This surprises many, as it sounds just a little too cold for comfort. Of course, that specific temperature will vary depending on the individual in question and their unique physiology, gender, and age. But like calorie recommendations, it's a good target for the average human being. Most of us set ambient house and/or bedroom temperatures higher than are optimal for good sleep and this likely contributes to lower quantity and/or quality of sleep than you are otherwise capable of getting. Lower than 55 degrees Fahrenheit (12.5°C) can be harmful rather than helpful to sleep, unless warm bedding or nightclothes are used. However, most of us fall into the opposite category of setting a controlled bedroom temperature that is too high: 70 or 72 degrees. Sleep clinicians treating insomnia patients will often ask about room temperature, and will advise patients to drop their current thermostat set-point by 3 to 5 degrees Fahrenheit from that which they currently use.

Anyone disbelieving of the influence of temperature on sleep can explore some truly bizarre experiments on this topic strewn throughout the research literature. Scientists have, for example, gently warmed the feet or the body of rats to encourage blood to rise to the surface of the skin and emit heat, thereby decreasing core body temperature. The rats drifted off to sleep far faster than was otherwise normal.

In a more outlandish human version of the experiment, scientists constructed a whole-body thermal sleeping suit, not dissimilar in appearance to a wet suit. Water was involved, but fortunately those willing to risk their dignity by donning the outfit did not get wet. Lining the suit was an intricate network of thin tubes, or veins. Crisscrossing the body like a detailed road map, these artificial veins traversed all major districts of the body: arms, hands, torso, legs, feet. And like the independent governance of local roads by separate states or counties of a nation, each body territory received its own separate water feed. In doing so, the scientists could exquisitely and selectively choose which parts of the body they would circulate water around, thereby controlling the temperature on the skin's surface in individual body areas—all while the participant lay quietly in bed.

Selectively warming the feet and hands by just a small amount (1°F, or about 0.5°C) caused a local swell of blood to these regions, thereby charming heat out of the body's core, where it had been trapped. The result of all this ingenuity: sleep took hold of the participants in a significantly shorter time, allowing them to fall asleep 20 percent faster than was usual, even though these were already young, healthy, fast-sleeping individuals.*

Not satisfied with their success, the scientists took on the challenge of improving sleep in two far more problematic groups: older adults who generally have a harder time falling asleep, and patients with clinical insomnia whose sleep was especially stubborn. Just like the young adults, the older adults fell asleep 18 percent faster than normal when receiving the same thermal assistance from the bodysuit. The improvement in the insomniacs was even more impressive—a 25 percent reduction in the time it took them to drift off into sleep.

Better still, as the researchers continued to apply body-temperature cooling throughout the night, the amount of time spent in stable sleep increased while time awake decreased. Before the body-cooling therapy, these groups had a 58 percent probability of waking up in the last half of the night and struggled to get back to sleep—a classic hallmark of sleep maintenance insomnia. This number tumbled to just a 4 percent likelihood when receiving thermal help from the bodysuit. Even the electrical quality of sleep—especially the deep, powerful brainwaves of NREM sleep—had been boosted by the thermal manipulation in all these individuals.

Knowingly or not, you have probably used this proven temperature manipulation to help your own sleep. A luxury for many is to draw a hot bath in the evening and soak the body before bedtime. We feel it helps us fall asleep more quickly, which it can, but for the opposite reason most people imagine. You do not fall asleep faster because you are toasty and warm to the core. Instead, the hot bath invites blood to the surface of your skin, giving you that flushed appearance. When you get out of the bath, those dilated blood vessels on the surface quickly help radiate out inner heat, and your core body temperature plummets. Consequently, you fall asleep more quickly because your core is colder. Hot baths prior to bed can also induce 10 to 15 percent more deep NREM sleep in healthy adults.†

---

*R. J. Raymann and Van Someren, "Diminished capability to recognize the optimal temperature for sleep initiation may contribute to poor sleep in elderly people," *Sleep* 31, no. 9 (2008): 1301–9.

†J. A. Horne and B. S. Shackell, "Slow wave sleep elevations after body heating: proxim-

AN ALARMING FACT

Adding to the impact of evening light and constant temperature, the industrial era inflicted another change to our sleep: enforced awakening. With the dawn of the industrial age and the emergence of large factories came a challenge: How can you guarantee the en masse arrival of a large workforce all at the same time, such as at the start of a shift?

The solution came in the form of the factory whistle—arguably the earliest (and loudest) version of an alarm clock. The whistle's skirl across the working village aimed to wrench large numbers of individuals from sleep at the same morning hour day after day. A second whistle would often signal the beginning of the work shift itself. Later, this invasive messenger of wakefulness entered the bedroom in the form of the modern-day alarm clock (and the second whistle was replaced by the banality of time card punching).

No other species demonstrates this unnatural act of prematurely and artificially terminating sleep.* Compare the physiological state of the body after being loudly awakened by an alarm to that observed after naturally waking from sleep. Participants artificially awakened from sleep can experience a spike in blood pressure and an acceleration in heart rate caused by a burst of activity from the fight-or-flight branch of the nervous system.†

Most of us are unaware of another consequence of the alarm clock: the snooze button. The snooze feature means that you will repeatedly impose that cardiovascular spike again and again within a short span of time. Step and repeat this at least five days a week, and you begin to understand the possible consequences to your heart and nervous system across a life span. Waking up at the same time of day, every day, no matter if it is the week or weekend is a good recommendation for main-

---

ity to sleep and effects of aspirin," *Sleep* 10, no. 4 (1987): 383–92. Also J. A. Horne and A. J. Reid, "Night-time sleep EEG changes following body heating in a warm bath," *Electroencephalography and Clinical Neurophysiology* 60, no. 2 (1985): 154–57.

*Not even roosters, since they crow not only at dawn but throughout the entire day.

†K. Kaida, K. Ogawa, M. Hayashi, and T. Hori, "Self-awakening prevents acute rise in blood pressure and heart rate at the time of awakening in elderly people," *Industrial Health* 43, no. 1 (January 2005): 179–85.

taining a stable sleep schedule if you are having difficulty with sleep. Indeed, it is one of the most consistent and effective ways of helping people with insomnia get better sleep. This unavoidably means the use of an alarm clock for many individuals, and that is understandable. If you do use an alarm clock, try to do away with the snooze function, and get in the habit of waking up only once to spare your heart the repeated shock.

Parenthetically, a hobby of mine is to collect the most innovative (i.e., ludicrous) alarm clock designs in some hope of cataloging the depraved ways we humans wrench our brains out of sleep. One such clock has a number of geometric blocks that sit in complementary-shaped holes on a pad. When the alarm goes off in the morning, it not only erupts into a blurting shriek, but also explodes the blocks out across the bedroom floor. It will not shut off the alarm until you pick up and reposition all of the blocks in their respective holes.

My favorite, however, is the shredder. You take a paper bill—let's say $20—and slide it into the front of the clock at night. When the alarm goes off in the morning, you have a short amount of time to wake up and turn the alarm off before it begins shredding your money. The brilliant behavioral economist Dan Ariely has suggested an even more fiendish system wherein your alarm clock is connected, by Wi-Fi, to your bank account. For every second you remain asleep, the alarm clock will send $10 to a political organization . . . that you absolutely despise.

That we have devised such creative—and even painful—ways of waking ourselves up in the morning says everything about how under-slept our modern brains are. Squeezed by the vise grips of an electrified night and early-morning start times, bereft of twenty-four-hour thermal cycles, and with caffeine and alcohol that some individuals may use in various quantities, many of us can feel understandably exhausted and crave that which seems always elusive: a full, restful night of natural deep sleep. The internal and external environments in which we evolved are not those in which we lie down to rest in the twenty-first century. To morph an agricultural concept from the wonderful writer and poet Wendell Berry,* modern society has taken one of nature's perfect solutions (sleep) and, for some people, neatly divided it into two problems: (1) a lack thereof at night, resulting in (2) an inability to remain fully awake during the day. These problems have forced some individuals to go in search of prescription sleeping pills. Is this wise? In the next chapter, I will provide you with information that speaks to this issue.

*"The genius of American farm experts is very well demonstrated here: they can take a solution and divide it neatly into two problems." From Wendell Berry, *The Unsettling of America: Culture & Agriculture* (1996), p. 62.

CHAPTER 14

Hurting and Helping Your Sleep

Pills vs. Therapy

In the past month, almost 10 million people in America will have swallowed some kind of a sleeping aid. Most relevant, and a key focus of this chapter, is the (ab)use of prescription sleeping pills. Sleeping pills do not provide natural sleep, can damage health, and increase the risk of life-threatening diseases. We will explore the alternatives that exist for improving sleep and combating insipid insomnia.

SHOULD YOU TAKE TWO OF THESE BEFORE BED?

No past or current sleeping medications on the legal (or illegal) market induce natural sleep. Don't get me wrong—no one would claim that you are awake after taking prescription sleeping pills. But to suggest that you are experiencing natural sleep would not be a true assertion.

The older sleep medications—termed "sedative hypnotics," such as diazepam—were blunt instruments. They sedated you rather than assisting you into natural sleep. Understandably, many people mistake the former for the latter. Most of the newer sleeping pills on the market present a similar situation, though they are slightly less heavy in their sedating effects. Sleeping pills, old and new, target the same system in the brain that alcohol does—the receptors that stop your brain cells from firing—and are thus part of the same general class of drugs: sedatives. Sleeping pills effectively knock out the higher regions of your brain's cortex.

If you compare natural, deep-sleep brainwave activity to that induced by modern-day sleeping pills, such as zolpidem (brand name Ambien) or eszopiclone (brand name Lunesta), the electrical signature, or quality, is deficient. The electrical type of "sleep" these drugs produce is lacking in the largest, deepest brainwaves.* Adding to this state of affairs are a number of unwanted side effects, including next-day grogginess, daytime forgetfulness, performing actions at night of which you are not conscious (or at least have partial amnesia of in the morning), and slowed reaction times during the day that can impact motor skills, such as driving.

True even of the newer, shorter-acting sleeping pills on the market, these symptoms may instigate a vicious cycle. The waking grogginess can lead people to reach for more cups of coffee or tea to rev themselves up with caffeine throughout the day and evening. That caffeine, in turn, makes it harder for the individual to fall asleep at night, worsening the insomnia. In response, people often take an extra half or whole sleeping pill at night to combat the caffeine, but this only amplifies the next-day grogginess from the drug hangover. Even greater caffeine consumption then occurs, perpetuating a possible downward spiral.

Another unpleasant feature of sleeping pills is rebound insomnia. When individuals stop taking these medications, they can suffer worse sleep, sometimes even worse than the poor sleep that led them to seek out sleeping pills to begin with. The cause of rebound insomnia is a type of dependency in which the brain alters its balance of receptors as a reaction to the increased drug dose, trying to become somewhat less sensitive as a way of countering the foreign chemical within the brain. This is also known as drug tolerance. But when the drug is stopped, there can be a withdrawal process, part of which involves an unpleasant spike in insomnia severity.

We should not be surprised by this. The majority of prescription sleeping pills are, after all, in a class of physically addictive drugs. Dependency scales with continued use, and withdrawal ensues in abstinence. Of course, when patients come off the drug for a night and suffer from poor sleep as a result of rebound insomnia, they often go right back to taking the drug. Few people realize that this night of

*E. L. Arbon, M. Knurowska, and D. J. Dijk, "Randomised clinical trial of the effects of prolonged release melatonin, temazepam and zolpidem on slow-wave activity during sleep in healthy people," *Journal of Psychopharmacology* 29, no. 7 (2015): 764–76.

severe insomnia, and the need to start retaking the drug, is partially or wholly caused by the persistent use of sleeping pills to begin with.

The irony is that many individuals experience only a modest increase in "sleep" from these medications, and the benefit can be more subjective than objective. A recent team of leading medical doctors and researchers examined all published studies to date on newer forms of sedative sleeping pills that most people take.* They considered sixty-five separate drug-placebo studies, encompassing almost 4,500 individuals. Overall, participants subjectively felt they fell asleep faster and slept more soundly with fewer awakenings, relative to the placebo. But that's not what the actual sleep recordings showed. There was no difference in how soundly the individuals slept. Both the placebo and the sleeping pills reduced the time it took people to fall asleep (between ten and thirty minutes), but the change was not statistically different between the two. In other words, there was no objective benefit of these sleeping pills beyond that which a placebo offered.

Summarizing the findings, the committee stated that sleeping pills only produced "slight improvements in subjective and polysomnographic sleep latency"—that is, the time it takes to fall asleep. The committee concluded the report by stating that the effect of current sleeping medications was "rather small and of questionable clinical importance." Even the newest sleeping pill for insomnia, called suvorexant (brand name Belsomra), has not proven to be as effective as hoped, as we discussed in chapter 12. Future versions of such drugs may offer meaningful sleep improvements, but for now the scientific data on prescription sleeping pills suggests that they may not be the answer to sound sleep to those struggling to generate it on their own, at least not as a long-term solution.

---

*T. B. Huedo-Medina, I. Kirsch, J. Middlemass, et al., "Effectiveness of non-benzodiazepine hypnotics in treatment of adult insomnia: meta-analysis of data submitted to the Food and Drug Administration," *BMJ* 345 (2012): e8343.

SLEEPING PILLS—THE BAD, THE BAD, AND THE UGLY

Existing prescription sleeping pills are minimally helpful, but are they associated with a higher risk of certain medical conditions, or even mortality?

Natural deep sleep, as we have previously learned, helps cement new memory traces within the brain, part of which require the active strengthening of connections between synapses that make up a memory circuit. How this essential nighttime storage function is affected by drug-induced sleep has been the focus of recent animal studies. After a period of intense learning, researchers at the University of Pennsylvania gave animals a weight-appropriate dose of Ambien or a placebo and then examined the change in brain rewiring after sleep in both groups. As expected, natural sleep solidified memory connections within the brain in the placebo condition that had been formed during the initial learning phase. Ambien-induced sleep, however, not only failed to match these benefits (despite the animals sleeping just as long), but caused a 50 percent *weakening* (unwiring) of the brain-cell connections originally formed during learning. In doing so, Ambien-laced sleep became a memory eraser, rather than engraver.

This is of special concern considering the average age for those receiving sleep medication prescriptions may be decreasing, as sleep complaints and incidents of pediatric insomnia increase. Should the former be true, doctors and parents may need to be vigilant about giving in to the temptation of prescriptions. Otherwise, young brains, which are still being wired up into the early twenties, will be attempting the already challenging task of neural development and learning under the influence of prescription sleeping pills.*

Perhaps most controversial are studies highlighted by Dr. Daniel

---

*A related concern is that of sleeping pill use in pregnant women. A recent scientific review of Ambien from a team of leading world experts stated: "[the] use of zolpidem [Ambien] should be avoided during pregnancy. It is believed that infants born to mothers taking sedative-hypnotic drugs such as zolpidem [Ambien] may be at risk for physical dependence and withdrawal symptoms during the postnatal period." (J. MacFarlane, C. M. Morin, and J. Montplaisir, "Hypnotics in insomnia: the experience of zolpidem," *Clinical Therapeutics* 36, no. 11 (2014): 1676–1701.)

Kripke, a physician at the University of California, San Diego. Kripke discovered that individuals using prescription sleep medications are significantly more likely to die and to develop cancer than those who do not.* I should note at the outset that Kripke (like me) has no vested interest in any particular drug company, and therefore does not stand to financially gain or lose on the basis of a particular examination of health relationships that exist with sleeping pills—good or bad.

In the early 2000s, insomnia rates ballooned and sleeping pill prescriptions escalated dramatically. It also meant much more data was available. Kripke began examining these large epidemiological databases. He wanted to explore whether there was a relationship between sleeping pill use and altered disease or mortality risk. There was. Time and again, the same message emerged from the analyses: individuals taking sleeping pills were significantly more likely to die across the study periods (usually a handful of years) than those who were not, the reasons for which we will soon discuss.

Often, however, it was tricky to conduct a well-matched comparison with these early databases, as there were not enough participants or measured factors that he could control for to really tease out a pure sleeping pill effect. By 2012, however, there were. Kripke and his colleagues set up a well-controlled comparison, examining more than 10,000 patients taking sleeping pills, the vast majority of whom were taking zolpidem (brand name Ambien), though some were taking temazepam (brand name Restoril). He contrasted them with 20,000 very well matched individuals of similar age, race, gender, and background, but who were not taking sleeping pills. In addition, Kripke was able to control for many other factors that could inadvertently contribute to mortality, such as body mass index, history of exercise, smoking, and drinking. He looked at the likelihood of disease and death across a two-and-a-half-year window, shown in Figure 15.†,‡

---

*D. F. Kripke, R. D. Langer, and L. E. Kline, "Hypnotics' association with mortality or cancer: a matched cohort study," *BMJ Open* 2, no. 1 (2012): e000850.

†Ibid.

‡Source: Dr. Daniel F. Kripke, "The Dark Side of Sleeping Pills: Mortality and Cancer Risks, Which Pills to Avoid & Better Alternatives," March 2013, accessed at http://www.darksideofsleepingpills.com.

Those taking sleeping pills were 4.6 times more likely to die over this short two-and-a-half-year period than those who were not using sleeping pills. Kripke further discovered that the risk of death scaled with the frequency of use. Those individuals classified as heavy users, defined as taking more than 132 pills per year, were 5.3 times more likely to die over the study period than matched control participants who were not using sleeping pills.

Figure 15: Risk of Death from Sleeping Pills

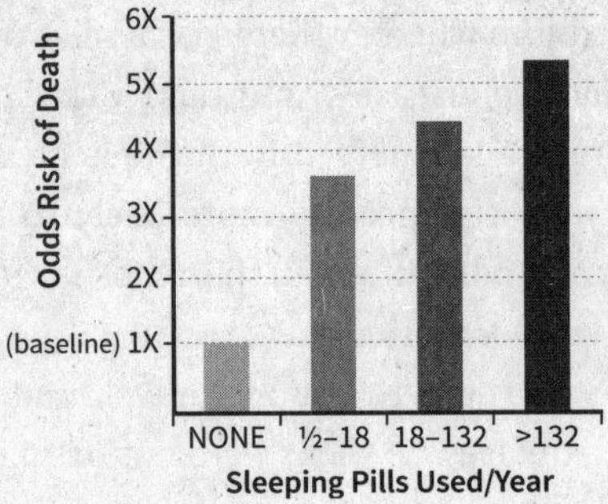

Even very occasional users—those defined as taking just eighteen pills per year—were still 3.6 times more likely to die at some point across the assessment window than non-users. Kripke isn't the only researcher finding such mortality risk associations. There are now more than fifteen such studies from different groups around the world showing higher rates of mortality in those who use sleeping pills.

What was killing those individuals using sleeping pills? That question is harder to answer from the available data, though it is clear that the sources could be many. In an attempt to find answers, Kripke and other independent research groups have now evaluated data from studies involving almost all of the common sleeping pills, including zolpidem (Ambien), temazepam (Restoril), eszopiclone (Lunesta), zaleplon (Sonata), and other sedating drugs, such as triazolam (Halcion) and flurazepam (Dalmane).

One frequent link with mortality appears to be higher-than-normal rates of infection. Also discussed in earlier chapters, natural sleep is one of the most powerful boosters of the immune system, helping ward off infection. Why, then, do individuals who are taking sleeping pills that purportedly "improve" sleep suffer *higher* rates of various infections, when the opposite is predicted? It is possible that medication-induced sleep does not provide the same restorative immune benefits as natural sleep. This would be troubling for the elderly. Older adults are far more likely to suffer from infections. Alongside newborns, they are the most immunologically vulnerable individuals in our society. Older adults are also the heaviest users of sleeping pills, representing more than 50 percent of the individuals prescribed such drugs. Based on these coincidental facts, it may be time for medicine to reappraise the prescription frequency of sleeping pills in the elderly.

Another possible concern of death linked to sleeping pill use is an increased risk for fatal car accidents. This could be due to the nonrestorative sleep such drugs induce and/or the groggy hangover that some suffer, both of which may leave individuals drowsy while driving the next day. Higher risk for falls at night was a further mortality factor, particularly in the elderly. Additional adverse associations in users of prescription sleeping pills included higher rates of heart disease and stroke.

Another topic of concern has been cancer. Earlier studies had hinted at a relationship between the sleep medications and mortality risk from cancer, but were not as well controlled in terms of comparisons. Kripke's study did a better job in this regard, and included the newer, more relevant sleeping medication Ambien. Individuals taking sleeping pills were 30 to 40 percent more likely to develop cancer within the two-and-a-half-year period of the study than those who were not. The older sleeping medications, such as temazepam (Restoril), had a stronger association, with those on mild to moderate doses suffering more than a 60 percent increased cancer risk. Those taking the highest dose of zolpidem (Ambien) were still vulnerable, suffering almost a 30 percent greater likelihood of developing cancer across the two-and-a-half-year study duration.

Do these findings prove that sleeping pills cause cancer? No. Absolutely not. There are alternative explanations. For example, it could be that the poor sleep that individuals were suffering prior to taking these drugs—that which motivated the prescription to begin with—and not the sleeping pills themselves, predisposed them to ill health. Moreover, the more problematic an individual's prior sleep, perhaps the more sleeping pills they later consumed, thus accounting for the dose-dependent mortality and dose-carcinogen relationships Kripke and others observed.

But it is equally possible that sleeping pills do cause death and cancer. To obtain a definitive answer we would need a dedicated clinical trial expressly designed to examine these particular morbidity and mortality risks.

Perhaps the most conservative and least litigious conclusion one can make about all of this evidence is that no study to date has truly shown that sleeping pills save lives. In my scientific, *though non-medical*, opinion, I believe that the existing evidence warrants more transparent medical education of any patient who is considering taking a sleeping pill, at the very least. This way, individuals can appreciate the risks and make informed choices. Do you, for example, feel differently about using or continuing to use sleeping pills having learned about this evidence?

To be very clear, I am not anti-medication. On the contrary, I desperately want there to be a drug that helps people obtain truly naturalistic sleep. Many of the drug company scientists who create sleeping medicines do so with nothing but good intent and an honest desire to help those for whom sleep is problematic. I know, because I have met many of them in my career. And as a researcher, I am keen to help science explore new medications in carefully controlled, independent studies. If such a drug—one with sound scientific data demonstrating benefits that far outweigh any health risks—is ultimately developed, I would support it. It is simply that no such medication currently exists.

DON'T TAKE TWO OF THESE, INSTEAD TRY THESE

While the search for more sophisticated sleep drugs continues, a new wave of exciting, non-pharmacological methods for improving sleep are fast emerging. Beyond the electrical, magnetic, and auditory stimulation methods for boosting deep-sleep quality that I have previously discussed (and that are still in embryonic stages of development) there are already numerous and effective behavioral methods for improving your sleep, especially if you are suffering from insomnia.

Currently, the most effective of these is called cognitive behavioral therapy for insomnia, or CBT-I, and it is rapidly being embraced by the medical community as the first-line treatment. Working with a therapist for several weeks, patients are provided with a bespoke set of techniques intended to break bad sleep habits and address anxieties that have been inhibiting sleep. CBT-I builds on basic sleep hygiene principles that I describe in the appendix, supplemented with methods individualized for the patient, their problems, and their lifestyle. Some are obvious, others not so obvious, and still others are counterintuitive.

The obvious methods involve reducing caffeine and alcohol intake, removing screen technology from the bedroom, and having a cool bedroom. In addition, patients must (1) establish a regular bedtime and wake-up time, even on weekends, (2) go to bed only when sleepy and avoid sleeping on the couch early/mid-evenings, (3) never lie awake in bed for a significant time period; rather, get out of bed and do something quiet and relaxing until the urge to sleep returns, (4) avoid daytime napping if you are having difficulty sleeping at night, (5) reduce anxiety-provoking thoughts and worries by learning to mentally decelerate before bed, and (6) remove visible clockfaces from view in the bedroom, preventing clock-watching anxiety at night.

One of the more paradoxical CBT-I methods used to help insomniacs sleep is to restrict their time spent in bed, perhaps even to just six hours of sleep or less to begin with. By keeping patients awake for longer, we build up a strong sleep pressure—a greater abundance of adenosine. Under this heavier weight of sleep pressure, patients fall asleep faster, and achieve a more stable, solid form of sleep across the night. In this way, a patient can regain their psychological confidence in being able to self-generate and sustain healthy, rapid, and sound sleep, night after night: something that has eluded them for months if not years. Upon reestablishing a patient's confidence in this regard, time in bed is gradually increased.

While this may all sound a little contrived or even dubious, skeptical readers, or those normally inclined toward drugs for help, should first evaluate the proven benefits of CBT-I before dismissing it outright. Results, which have now been replicated in numerous clinical studies around the globe, demonstrate that CBT-I is more effective than sleeping pills in addressing numerous problematic aspects of sleep for insomnia sufferers. CBT-I consistently helps people fall asleep faster at night, sleep longer, and obtain superior sleep quality by significantly decreasing the amount of time spent awake at night.* More importantly, the benefits of CBT-I persist long term, even after patients stop working with their sleep therapist. This sustainability stands in stark contrast to the punch of rebound insomnia that individuals experience following the cessation of sleeping pills.

So powerful is the evidence favoring CBT-I over sleeping pills for improved sleep across all levels, and so limited or nonexistent are the safety risks associated with CBT-I (unlike sleeping pills), that in 2016, the American College of Physicians made a landmark recommendation. A committee of distinguished sleep doctors and scientists evaluated all aspects of the efficacy and safety of CBT-I relative to standard sleeping pills. Published in the prestigious journal *Annals of Internal Medicine*, the conclusion from this comprehensive evaluation of all existing data was this: CBT-I must be used as *the* first-line treatment for all individuals with chronic insomnia, not sleeping pills.†

You can find more resources on CBT-I, and a list of qualified therapists, from the National Sleep Foundation's website.‡ If you have, or

---

*M. T. Smith, M. L. Perlis, A. Park, et al., "Comparative meta-analysis of pharmacotherapy and behavior therapy for persistent insomnia," *American Journal of Psychiatry* 159, no. 1 (2002): 5–11.

†Such committees will also assign a weighted grade to their clinical recommendation, from mild to moderate to strong. This grade helps guide and inform GPs across the nation regarding how judiciously they should apply the ruling. The committee's grading on CBT-I was: Strongly Recommend.

‡https://sleepfoundation.org.

think you have, insomnia, please make use of these resources before turning to sleeping pills.

GENERAL GOOD SLEEP PRACTICES

For those of us who are not suffering from insomnia or another sleep disorder, there are things we can do to secure a far better night of sleep using what we call good "sleep hygiene" practices, for which a list of twelve key tips can be found at the National Institutes of Health website; also offered in the appendix of this book.* All twelve suggestions are superb advice, but if you can only adhere to one of these each and every day, make it: going to bed and waking up at the same time of day no matter what. It is perhaps the single most effective way of helping improve your sleep, even though it involves the use of an alarm clock.

Last but not least, two of the most frequent questions I receive from members of the public regarding sleep betterment concern exercise and diet.

Sleep and physical exertion have a bidirectional relationship. Many of us know of the deep, sound sleep we often experience after sustained physical activity, such as a daylong hike, an extended bike ride, or even an exhausting day of working in the garden. Scientific studies dating back to the 1970s support some of this subjective wisdom, though perhaps not as strongly as you'd hope. One such early study, published in 1975, shows that progressively increased levels of physical activity in healthy males results in a corresponding progressive increase in the amount of deep NREM sleep they obtain on subsequent nights. In another study, however, active runners were compared with age- and gender-matched non-runners. While runners had somewhat higher amounts of deep NREM sleep, it was not significantly different to the non-runners.

Larger and more carefully controlled studies offer somewhat more positive news, but with an interesting wrinkle. In younger, healthy

*"Tips for Getting a Good Night's Sleep," *NIH Medline Plus*. Accessed at https://www.nlm.nih.gov/medlineplus/magazine/issues/summer12/articles/summer12pg20.html (or just search the Internet for "12 tips for better sleep, NIH").

adults, exercise frequently increases total sleep time, especially deep NREM sleep. It also deepens the quality of sleep, resulting in more powerful electrical brainwave activity. Similar, if not larger, improvements in sleep time and efficiency are to be found in midlife and older adults, including those who are self-reported poor sleepers or those with clinically diagnosed insomnia.

Typically, these studies involve measuring several nights of initial baseline sleep in individuals, after which they are placed on a regimen of exercise across several months. Researchers then examine whether or not there are corresponding improvements in sleep as a consequence. On average, there are. Subjective sleep quality improves, as does total amount of sleep. Moreover, the time it takes participants to fall asleep is usually less, and they report waking up fewer times across the night. In one of the longest manipulation studies to date, older adult insomniacs were sleeping almost one hour more each night, on average, by the end of a four-month period of increased physical activity.

Unexpected, however, was the lack of a tight relationship between exercise and subsequent sleep from one day to the next. That is, subjects did not consistently sleep better at night on the days they exercised compared with the days when they were not required to exercise, as one would expect. Less surprising, perhaps, is the inverse relationship between sleep and *next-day* exercise (rather than the influence of exercise on subsequent sleep at night). When sleep was poor the night prior, exercise intensity and duration were worse the following day. When sleep was sound, levels of physical exertion were powerfully maximal the next day. In other words, sleep may have more of an influence on exercise than exercise has on sleep.

It is still a clear bidirectional relationship, however, with a significant trend toward increasingly better sleep with increasing levels of physical activity, and a strong influence of sleep on daytime physical activity. Participants also feel more alert and energetic as a result of the sleep improvement, and signs of depression proportionally decrease. It is clear that a sedentary life is one that does not help with sound sleep, and all of us should try to engage in some degree of regular exercise to help maintain not only the fitness of our bodies but also the quantity and quality of our sleep. Sleep, in return, will boost your fitness and energy, setting in motion a positive, self-sustaining cycle of improved physical activity (and mental health).

One brief note of caution regarding physical activity: try not to exercise right before bed. Body temperature can remain high for an hour or two after physical exertion. Should this occur too close to bedtime, it can be difficult to drop your core temperature sufficiently to initiate sleep due to the exercise-driven increase in metabolic rate. Best to get your workout in at least a few hours before turning the bedside light out (none LED-powered, I trust).

When it comes to diet, there is limited research investigating how the foods you eat, and the pattern of eating, impact your sleep at night. Severe caloric restriction, such as reducing food intake to just 800 calories a day for one month, makes it harder to fall asleep normally, and decreases the amount of deep NREM sleep at night.

What you eat also appears to have some impact on your nighttime sleep. Eating a high-carbohydrate, low-fat diet for two days decreases the amount of deep NREM sleep at night, but increases the amount of REM sleep dreaming, relative to a two-day diet low in carbohydrates and high in fat. In a carefully controlled study of healthy adult individuals, a four-day diet high in sugar and other carbohydrates, but low in fiber, resulted in less deep NREM sleep and more awakenings at night.*

It is hard to make definitive recommendations for the average adult, especially because larger-scale epidemiological studies have not shown consistent associations between eating specific food groups and sleep quantity or quality. Nevertheless, for healthy sleep, the scientific evidence suggests that you should avoid going to bed too full or too hungry, and shy away from diets that are excessively biased toward carbohydrates (greater than 70 percent of all energy intake), especially sugar.

*M. P. St-Onge, A. Roberts, A. Shechter, and A. R. Choudhury, "Fiber and saturated fat are associated with sleep arousals and slow wave sleep," *Journal of Clinical Sleep Medicine* 12 (2016): 19–24.

CHAPTER 15

Sleep and Society:

What Medicine and Education Are Doing Wrong; What Google and NASA Are Doing Right

Back in 1942, a survey indicated that just 8 percent of the population of the United States slept six hours or less a night. Now, almost 25 percent of American adults do.

The lens of a 2013 survey by the National Sleep Foundation pulls this sleep deficiency into sharp focus.* Almost 35 percent of the US adult population fail to obtain the CDC-recommended minimum of seven hours of sleep a night. Circumnavigate the globe, and things look no better. In the UK and Japan, for example, 39 and 66 percent, respectively, of all adults report sleeping fewer than seven hours. Deep currents of sleep neglect circulate throughout all developed nations, and it is for these reasons that the World Health Organization now labels the lack of societal sleep as a global health epidemic. Taken as a whole, one out of every two adults across all developed countries (approximately 800 million people) will not get the necessary sleep they need this coming week.

Importantly, many of these individuals do not report *wanting* or *needing* less sleep. If you look at sleep time in first-world nations for the weekends, the numbers are very different. Rather than 30 percent of adults getting eight hours of sleep or more on average, almost 60 percent of these individuals attempt to "binge" on eight or more hours. Each weekend, substantial numbers of people are trying to pay back a sleep debt they've accrued during the week. As we have learned time

* National Sleep Foundation, 2013 International Bedroom Poll, accessed at https://sleepfoundation.org/sleep-polls-data/other-polls/2013-international-bedroom-poll.

and again throughout the course of this book, sleep is not like a credit system or the bank. The brain can never recover all the sleep it has been deprived of. We cannot accumulate a debt without penalty, nor can we repay that sleep debt in full at a later time.

Beyond any single individual, why should society care? Would altering sleep attitudes and increasing sleep amounts make any difference to our collective lives as a human race, to our professions and corporations, to commercial productivity, to salaries, the education of our children, or even our moral nature? Whether you are a business leader or employee, the director of a hospital, a practicing doctor or nurse, a government official or military person, a public-policy maker or community health worker, anyone who expects to receive any form of medical care at any moment in their life, or a parent, the answer is very much "yes," for more reasons than you may imagine.

Below, I offer four diverse yet clear examples of how insufficient sleep is impacting the fabric of human society. These are: sleep in the workplace, torture (yes, torture), sleep in the education system, and sleep in medicine and health care.

SLEEP IN THE WORKPLACE

Sleep deprivation degrades many of the key faculties required for most forms of employment. Why, then, do we overvalue employees that undervalue sleep? We glorify the high-powered executive on email until 1:00 a.m., and then in the office by 5:45 a.m.; we laud the airport "warrior" who has traveled through five different time zones on seven flights over the past eight days.

There remains a contrived, yet fortified, arrogance in many business cultures focused on the uselessness of sleep. It is bizarre, considering how sensible the professional world is regarding all other areas of employee health, safety, and conduct. As my Harvard colleague, Dr. Czeisler has pointed out, innumerable policies exist within the workplace regarding smoking, substance abuse, ethical behavior, and injury and disease prevention. But insufficient sleep—another harmful, potentially deadly factor—is commonly tolerated and even woefully encouraged. This mentality has persisted, in part, because certain business leaders mistakenly believe that time on-task equates with task completion and productivity. Even in the industrial era of rote factory work, this was untrue. It is a misguided fallacy, and an expensive one, too.

A study across four large US companies found that insufficient sleep cost almost $2,000 per employee per year in lost productivity. That amount rose to over $3,500 per employee in those suffering the most serious lack of sleep. That may sound trivial, but speak to the bean counters that monitor such things and you discover a net capital loss to these companies of $54 million annually. Ask any board of directors whether they would like to correct a single problem fleecing their company of more than $50 million a year in lost revenue and the vote will be rapid and unanimous.

An independent report by the RAND Corporation on the economic cost of insufficient sleep offers a sobering wake-up call for CFOs and CEOs.* Individuals who sleep fewer than seven hours a night on average cause a staggering fiscal cost to their country, compared to employees who sleep more than eight hours each night. Shown in figure 16A, inadequate sleep costs America and Japan $411 billion and $138 billion each year, respectively. The UK, Canada, and Germany follow.

Figure 16: Global Economic Cost of Sleep Loss

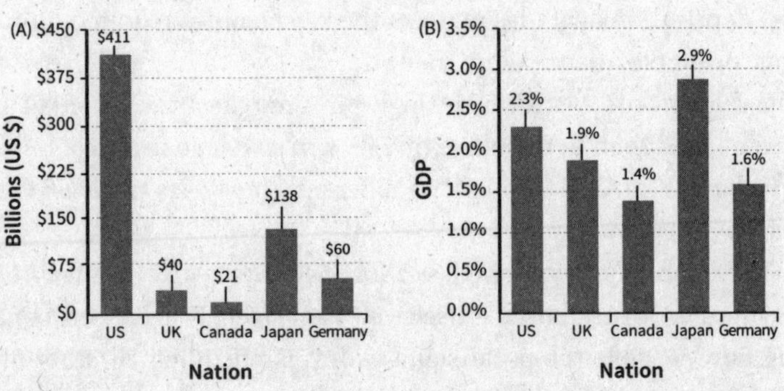

Of course, these numbers are skewed by the size of the country. A standardized way to appreciate the impact is by looking at gross

*"RAND Corporation, Lack of Sleep Costing UK Economy Up to £40 Billion a Year," accessed at http://www.rand.org/news/press/2016/11/30/index1.html.

domestic product (GDP)—a general measure of a country's profit output, or economic health. Viewed this way, things look even more bleak, described in figure 16B. Insufficient sleep robs most nations of more than 2 percent of their GDP—amounting to the entire cost of each country's military. It's almost as much as each country invests in education. Just think, if we eliminated the national sleep debt, we could almost double the GDP percentage that is devoted to the education of our children. One more way that abundant sleep makes financial sense, and should itself be incentivized at the national level.

Why are individuals so financially ruinous to their companies, and national economies, when they are under-slept? Many of the Fortune 500 companies that I give presentations to are interested in KPIs—key performance indicators, or measurables, such as net revenue, goal-accomplishment speed, or commercial success. Numerous employee traits determine these measures, but commonly they include: creativity, intelligence, motivation, effort, efficiency, effectiveness when working in groups, as well as emotional stability, sociability, and honesty. All of these are systematically dismantled by insufficient sleep.

Early studies demonstrated that shorter sleep amounts predict lower work rate and slow completion speed of basic tasks. That is, sleepy employees are unproductive employees. Sleep-deprived individuals also generate fewer and less accurate solutions to work-relevant problems they are challenged with.*

We have since designed more work-relevant tasks to explore the effects of insufficient sleep on employee effort, productivity, and creativity. Creativity is, after all, lauded as the engine of business innovation. Give participants the ability to choose between work tasks of varying effort, from easy (e.g., listening to voice mails) to difficult (e.g., helping design a complex project that requires thoughtful problem solving and creative planning), and you find that those individuals who obtained less sleep in the preceding days are the same people who consistently select less challenging problems. They opt for the easy way out, generating fewer creative solutions in the process.

---

*W. B. Webb and C. M. Levy, "Effects of spaced and repeated total sleep deprivation," *Ergonomics* 27, no. 1 (1984): 45–58.

It is, of course, possible that the type of people who decide to sleep less are also those who prefer not to be challenged, and one has nothing directly to do with the other. Association does not prove causation. However, take the same individuals and repeat this type of experiment twice, once when they have had a full night of sleep and once when they are sleep-deprived, and you see the same effects of laziness caused by a lack of sleep when using each person as their own baseline control.* A lack of sleep, then, is indeed a causal factor.

Under-slept employees are not, therefore, going to drive your business forward with productive innovation. Like a group of people on stationary exercise bikes, everyone looks like they are pedaling, but the scenery never changes. The irony that employees miss is that when you are not getting enough sleep, you work less productively and thus need to work longer to accomplish a goal. This means you often must work longer and later into the evening, arrive home later, go to bed later, and need to wake up earlier, creating a negative feedback loop. Why try to boil a pot of water on medium heat when you could do so in half the time on high? People often tell me that they do not have enough time to sleep because they have so much work to do. Without wanting to be combative in any way whatsoever, I respond by informing them that perhaps one reason they still have so much to do at the end of the day is perhaps because they do not get enough sleep at night.

Interestingly, participants in the above studies do not perceive themselves as applying less effort to the work challenge, or being less effective, when they were sleep-deprived, despite both being true. They seemed unaware of their poorer work effort and performance—a theme of subjective misperception of ability when sleep-deprived that we have touched upon previously in this book. Even the simplest daily routines that require slight effort, such as time spent dressing neatly or fashionably for the workplace, have been found to decrease following a night of sleep loss.† Individuals also like their jobs less when sleep-deprived—

*M. Engle-Friedman and S. Riela, "Self-imposed sleep loss, sleepiness, effort and performance," *Sleep and Hypnosis* 6, no. 4 (2004): 155–62; and M. Engle-Friedman, S. Riela, R. Golan, et al., "The effect of sleep loss on next day effort," *Journal of Sleep Research* 12, no. 2 (2003): 113–24.

†Ibid.

perhaps unsurprising considering the mood-depressing influence of sleep deficiency.

Under-slept employees are not only less productive, less motivated, less creative, less happy, and less effective, but they may be more unethical. Reputation in business can be a make-or-break factor. Having under-slept employees in your business makes you more vulnerable to that risk of disrepute. Previously, I described evidence from brain-scanning experiments showing that the frontal lobe, which is critical for self-control and reining in emotional impulses, is taken offline by a lack of sleep. As a result, participants were more emotionally volatile and rash in their choices and decision-making. This same result is predictably borne out in the higher-stakes setting of the workplace.

Studies in the workplace have found that employees who sleep six hours or less are significantly more deviant and more likely to lie the following day than those who sleep six hours or more. Seminal work by Dr. Christopher Barns, a researcher in the Foster School of Business at Washington University, has found that the less an individual sleeps, the more likely they are to create fake receipts and reimbursement claims, and the more willing to lie to get free raffle tickets. Barns also discovered that under-slept employees are more likely to blame other people in the workplace for their own mistakes, and even try to take credit for other people's successful work: hardly a recipe for team building and a harmonious business environment.

Ethical deviance linked to a lack of sleep also weasels its way onto the work stage in a different guise, called social loafing. The term refers to someone who, when group performance is being assessed, decides to exert less effort when working in that group than when working alone. Individuals see an opportunity to slack off and hide behind the collective hard work of others. They complete fewer aspects of the task themselves, and that work tends to be either wrong or of lower quality, relative to when they alone are being assessed. Sleepy employees therefore choose the more selfish path of least resistance when working in teams, coasting by on the disingenuous ticket of social loafing.* Not

---

*C. Y. Hoeksema-van Orden, A. W. Gaillard, and B. P. Buunk, "Social loafing under fatigue," *Journal of Personality and Social Psychology* 75, no. 5 (1998): 1179–90.

only does this lead to lower group productivity, understandably it often creates feelings of resentment and interpersonal aggression among team members.

Of note to those in business, many of these studies report deleterious effects on business outcomes on the basis of modest reductions in sleep amount within an individual, perhaps twenty- to sixty-minute differences between an employee who is honest, creative, innovate, collaborative, and productive and one who is not.

Examine the effects of sleep deficiency in CEOs and supervisors, and the story is equally impactful. An ineffective leader within any organization can have manifold trickle-down consequences to the many whom they influence. We often think that a good or bad leader is good or bad day after day—a stable trait. Not true. Differences in individual leadership performance fluctuate dramatically from one day to the next, and the size of that difference far exceeds the average difference from one individual leader to another. So what explains the ups and downs of a leader's ability to effectively lead, day to day? The amount of sleep they are getting is one clear factor.

A deceptively simple but clever study tracked the sleep of supervisors across several weeks, and compared that with their leadership performance in the workplace as judged by the employees who report to them. (I should note that employees themselves had no knowledge of how well their boss was sleeping each night, taking away any knowledge bias.) The lower the quality of sleep that the supervisor reported getting from one night to the next accurately predicted poor self-control and a more abusive nature toward employees the following day, as reported by the employees themselves.

There was another equally intriguing result: in the days after a supervisor had slept poorly, the employees themselves, even if well rested, became less engaged in their jobs throughout that day as a consequence. It was a chain-reaction effect, one in which the lack of sleep in that one superordinate person in a business structure was transmitted on like a virus, infecting even well-rested employees with work disengagement and reduced productivity.

Reinforcing this reciprocity, we have since discovered that underslept managers and CEOs are less charismatic and have a harder time infusing their subordinate teams with inspiration and drive. Unfortunately for bosses, a sleep-deprived employee will erroneously perceive a well-rested leader as being significantly less inspiring and charismatic than they truly are. One can imagine the multiplicative consequences to the success of a business if both the leader and the employees are overworked and under-slept.

Allowing and encouraging employees, supervisors, and executives to arrive at work well rested turns them from simply looking busy yet ineffective, to being productive, honest, useful individuals who inspire, support, and help each other. Ounces of sleep offer pounds of business in return.

Employees also win financially when sleep times increase. Those who sleep more earn more money, on average, as economists Matthew Gibson and Jeffrey Shrader discovered when analyzing workers and their pay across the United States. They examined townships of very similar socioeducational and professional standing within the same time zone, but at very far western and eastern edges of these zones that receive significantly different amounts of daylight hours. Workers in the far western locations obtained more sunlight later into the evening, and consequently went to bed an hour later, on average, than those in the far eastern locations. However, all workers in both regions had to wake up at the same time each morning, since they were all in the same time zone and on the same schedule. Therefore, western-dwelling workers in that time zone had less sleep opportunity time than the eastern-dwelling workers.

Factoring out many other potential factors and influences (e.g., regional affluence, house prices, cost of living, etc.), they found that an hour of extra sleep still returned significantly higher wages in those eastern locations, somewhere in the region of 4 to 5 percent. You may sniff at that return on the investment of sixty minutes of sleep, but it's not trivial. The average pay raise in the US is around 2.6 percent. Most people are strongly motivated to get that raise, and are upset when they don't. Imagine almost doubling that pay raise—not by working more hours, but by getting more sleep!

The fact of the matter is that most people will trade sleep for a higher salary. A recent study from Cornell University surveyed hundreds of US workers and gave them a choice between either (1) $80,000 a year, working normal work hours, and getting the chance for around eight hours of sleep, or (2) $140,000 a year, working consistent overtime shifts, and only getting six hours of sleep each night. Unfortunately, the majority of individuals went with the second option of a higher salary and shorter sleep. That's ironic, considering that you can have both, as we have discovered above.

The loud-and-proud corporate mentality of sleeplessness as the model for success is evidentially wrong at every level of analysis we have explored. Sound sleep is clearly sound business. Nevertheless, many companies remain somewhat antisleep in their structured practices. Like flies set in amber, this attitude keeps their businesses in a similarly frozen state of stagnation, lacking in innovation and productivity, and breeding employee unhappiness, dissatisfaction, and ill health.

There are, however, an increasing number of forward-looking companies who have changed their work practices in response to these research findings, and even welcome scientists like me into their businesses to teach and extol the virtues of getting more sleep to senior leaders and management. Procter & Gamble Co. and Goldman Sachs Group Inc., for example, both offer free "sleep hygiene" courses to their employees. Expensive, high-grade lighting has been installed in some of their buildings to better help workers regulate their circadian rhythms, improving the timed release of melatonin.

Nike and Google have both adopted a more relaxed approach to work schedules, allowing employees to time their daily work hours to match their individual circadian rhythms and their respective owl and lark chronotype nature. The change in mind-set is so radical that these same brand-leading corporations even allow workers to sleep on the job. Littered throughout their corporate headquarters are dedicated relaxation rooms with "nap pods." Employees can indulge in sleep throughout the workday in these "shh" zones, germinating productivity and creativity while enhancing wellness and reducing absenteeism.

Such changes reflect a marked departure from the draconian days when any employee found catnapping on the clock was chastised, disciplined, or outright fired. Sadly, most CEOs and managers still reject the importance of a well-slept employee. They believe such accommodations represent the "soft approach." But make no mistake: companies like Nike and Google are as shrewd as they are profitable. They embrace sleep due to its proven dollar value.

One organization above all has known about the occupational benefits of sleep longer than most. In the mid-1990s, NASA refined the science of sleeping on the job for the benefit of their astronauts. They discovered that naps as short as twenty-six minutes in length still offered a 34 percent improvement in task performance and more than a 50 percent increase in overall alertness. These results hatched the so-called NASA nap culture throughout terrestrial workers in the organization.

By any metrics we use to determine business success—profit margins, marketplace dominance/prominence, efficiency, employee creativity, or worker satisfaction and wellness—creating the necessary conditions for employees to obtain enough sleep at night, or in the workplace during the day, should be thought of as a new form of physiologically injected venture capital.

THE INHUMANE USE OF SLEEP LOSS IN SOCIETY

Business is not the only place where sleep deprivation and ethics collide. Governments and militaries bare a more disgraceful blemish.

Aghast at the mental and physical harm caused by prolonged sleep deprivation, in the 1980s Guinness ceased to recognize any attempts to break the world record for sleep deprivation. It is for similar reasons that scientists have limited evidence of the long-term effects of total sleep deprivation (beyond a night or two). We feel it morally unacceptable to impose that state on humans—and increasingly, on any species.

Some governments do not share these same moral values. They will sleep deprive individuals against their will under the auspice of torture. This ethically and politically treacherous landscape may seem like an odd topic to include in this book. But I address it because it powerfully illuminates how humanity must reevaluate its views on sleep at the highest level of societal structure—that of government—and because it provides a clear example of how we can sculpt an increasingly admirable civilization by respecting, rather than abusing, sleep.

A 2007 report entitled "Leave No Marks: Enhanced Interrogation Techniques and the Risk of Criminality" offers a disquieting account of such practices in the modern day. The document was compiled by Physicians for Human Rights, an advocacy group seeking to end human torture. Telegraphed by the report's title, many modern-day torture methods are deviously designed to leave no evidence of physical assault. Sleep deprivation epitomizes this goal and, at the time of writing this book, is still used for interrogation by countries, including Myanmar, Iran, Iraq, the United States, Israel, Egypt, Libya, Pakistan, Saudi Arabia, Tunisia, and Turkey.

As a scientist intimate with the workings of sleep, I would argue strongly for the abolition of this practice, structured around two clear facts. The first, and less important, is simply on grounds of pragmatism. In the context of interrogation, sleep deprivation is ill designed for the purpose of obtaining accurate, and thus actionable, intelligence. A lack of sleep, even moderate amounts, degrades every mental faculty necessary to obtain valid information, as we have seen. This includes the loss of accurate memory recall, emotional instability that prevents logical thought, and even basic verbal comprehension. Worse still, sleep deprivation increases deviant behavior and causes higher rates of lying and dishonesty.* Short of coma, sleep deprivation places an individual into the least useful brain state for the purpose of credible intelligence gathering: a disordered mind from which false confessions will flourish—which, of course, could be the intent of some captors. Proof comes from a recent scientific study demonstrating that one night of sleep deprivation will double or even quadruple the likelihood that an otherwise upstanding individual will falsely confess to something they have not done. You can, therefore, change someone's very attitudes, their behavior, and even their strongly held beliefs simply by taking sleep away from them.

An eloquent yet distressing affirmation of this fact is provided by the former prime minister of Israel, Menachem Begin, in his autobiography, *White Nights: The Story of a Prisoner in Russia.* In the 1940s, years

*C. M. Barnesa, J. Schaubroeckb, M. Huthc, and S. Ghummand, "Lack of sleep and unethical conduct," *Organizational Behavior and Human Decision Processes* 115, no. 3 (2011): 169–80.

before taking office in 1977, Begin was captured by the Soviets. He was tortured in prison by the KGB, one component of which involved prolonged sleep deprivation. Of this experience (which most governments benignly describe as the practice of "prisoner sleep management"), he writes:

> In the head of the interrogated prisoner a haze begins to form. His spirit is wearied to death, his legs are unsteady, and he has one sole desire: to sleep, to sleep just a little, not to get up, to lie, to rest, to forget . . . Anyone who has experienced this desire knows that not even hunger or thirst are comparable with it . . . I came across prisoners who signed what they were ordered to sign, only to get what their interrogator promised them. He did not promise them their liberty. He promised them—if they signed—uninterrupted sleep.

The second and more forceful argument for the abolition of enforced sleep deprivation is the permanent physical and mental harm it inflicts. Unfortunately, though conveniently for interrogators, the harm inflicted is not obvious from the outside. Mentally, long-term sleep deprivation over many days elevates suicidal thoughts and suicide attempts, both of which occur at vastly higher rates in detained prisoners relative to the general population. Inadequate sleep further cultivates the disabling and non-transient conditions of depression and anxiety. Physically, prolonged sleep deprivation increases the likelihood of a cardiovascular event, such as a heart attack or stroke, weakens the immune system in ways that encourage illness and disease.

Several US federal courts hold a similarly damning view of these practices, ruling that sleep deprivation violates both the Eighth and Fourteenth Amendments of the United States Constitution regarding protection from cruel and inhuman punishment. Their rationale was sound and impenetrable: "sleep," it was stated, must be considered a "basic life necessity," which it clearly is.

Nevertheless, the US Department of Defense subverted this ruling, authorizing twenty-hour interrogations of detainees in Guantánamo Bay between 2003 and 2004. Such treatment remains permissible to this day of writing, as the revised *US Army Field Manual* states, in appendix M, that detainees can be limited to just four hours of sleep every twenty-four hours, for up to four weeks. I note that it was not always so. A much earlier 1992 edition of the same publication held that extended sleep deprivation was a clear and inhumane example of "mental torture."

Depriving a human of sleep without their willing consent and careful medical care is a barbaric tool of assault, psychologically and biologically. Measured on the basis of mortality impact over the long term, it is on a par with starvation. It is high time to close the chapter on torture, including the use of sleep deprivation—an unacceptable and inhumane practice, one that I believe we will look back on with the very deepest of shame in years to come.

SLEEP AND EDUCATION

More than 80 percent of public high schools in the United States begin before 8:15 a.m. Almost 50 percent of those start before 7:20 a.m. School buses for a 7:20 a.m. start time usually begin picking up kids at around 5:45 a.m. As a result, some children and teenagers must wake up at 5:30 a.m., 5:15 a.m., or even earlier, and do so five days out of every seven, for years on end. This is lunacy.

Could you concentrate and learn much of anything when you had woken up so early? Keep in mind that 5:15 a.m. to a teenager is not the same as 5:15 a.m. to an adult. Previously, we noted that the circadian rhythm of teenagers shifts forward dramatically by one to three hours. So really the question I should ask you, if you are an adult, is this: Could you concentrate and learn anything after having forcefully been woken up at 3:15 a.m., day after day after day? Would you be in a cheerful mood? Would you find it easy to get along with your coworkers and conduct yourself with grace, tolerance, respect, and a pleasant demeanor? Of course not. Why, then, do we ask this of the millions of teenagers and children in industrialized nations? Surely this is not an optimal design of education. Nor does it bear any resemblance to a model for nurturing good physical or mental health in our children and teenagers.

Forced by the hand of early school start times, this state of chronic sleep deprivation is especially concerning considering that adolescence is the most susceptible phase of life for developing chronic mental illnesses, such as depression, anxiety, schizophrenia, and suicidality. Unnecessarily bankrupting the sleep of a teenager could make all the difference in the precarious tipping point between psychological wellness and lifelong psychiatric illness. This is a strong statement, and I do not write it flippantly or without evidence. Back in the 1960s, when the functions of sleep were still largely unknown, researchers selectively deprived young adults of REM sleep, and thus dreaming, for a week, while still allowing them NREM sleep.

The unfortunate study participants spent the entire time in the laboratory with electrodes placed on their heads. At night, whenever they entered into the REM-sleep state, a research assistant would quickly enter the bedroom and wake the subjects up. The blurry-eyed participants then had to do math problems for five to ten minutes, preventing them from falling back into dream sleep. But as soon as the participants did return into REM sleep, the procedure was repeated. Hour after hour, night after night, this went on for an entire week. NREM sleep was left largely intact, but the amount of REM sleep was reduced to a fraction of its regular quantity.

It didn't require all seven nights of dream-sleep deprivation before the mental health effects began to manifest. By the third day, participants were expressing signs of psychosis. They became anxious, moody, and started to hallucinate. They were hearing things and seeing things that were not real. They also became paranoid. Some believed that the researchers were plotting against them in collusive ways—trying to poison them, for example. Others became convinced that the scientists were secret agents, and that the experiment was a thinly veiled government conspiracy of some wicked kind.

Only then did scientists realize the rather profound conclusions of the experiment: REM sleep is what stands between rationality and insanity. Describe these symptoms to a psychiatrist without informing them of the REM-sleep deprivation context, and the clinician will give clear diagnoses of depression, anxiety disorders, and schizophrenia. But these were all healthy young individuals just days before. They were not depressed, weren't suffering from anxiety disorders or schizophrenia, nor did they have any history of such conditions, self or familial. Read of any attempts to break sleep-deprivation world records throughout early history, and you will discover this same universal signature of emotional instability and psychosis of one sort or another. It is the lack of REM sleep—that critical stage occurring in the final hours of sleep that we strip from our children and teenagers by way of early school start times—that creates the difference between a stable and unstable mental state.

Our children didn't always go to school at this biologically unreasonable time. A century ago, schools in the US started at nine a.m. As a result, 95 percent of all children woke up without an alarm clock. Now, the inverse is true, caused by the incessant marching back of school start times—which are in direct conflict with children's evolutionarily preprogrammed need to be asleep during these precious, REM-sleep-rich morning hours.

Growing scientific evidence now supports the wisdom of later school start times. One longitudinal study tracked more than 5,000 Japanese schoolchildren and discovered that those individuals who were sleeping longer obtained better grades across the board. Controlled sleep laboratory studies in smaller samples show that children with longer total sleep times develop superior IQ, with brighter children having consistently slept forty to fifty minutes more than those who went on to develop a lower IQ.

Examinations of identical twins further impress how powerful sleep is as a factor that can alter genetic determinism. In a study that was started by Dr. Ronald Wilson at Louisville School of Medicine in the 1980s, which continues to this day, hundreds of twin pairs were assessed at a very young age. The researchers specifically focused on those twins in which one was routinely obtaining less sleep than the other, and tracked their developmental progress over the following decades. By ten years of age, the twin with the longer sleep pattern was superior in their intellectual and educational abilities, with higher scores on standardized tests of reading and comprehension, and a more expansive vocabulary than the twin who was obtaining less sleep.

Such associational evidence is not proof that sleep is causing such powerful educational benefits. Nevertheless, combined with causal evidence linking sleep to memory that we have covered in chapter 6, a prediction can be made: if sleep really is so rudimentary to learning, then increasing sleep time by delaying start times should prove transformative. It has.

A growing number of schools in the US have started to revolt against the early start time model, beginning the school day at somewhat more biologically reasonable times. One of the first test cases happened in the township of Edina, Minnesota. Here, school start times for teenagers were shifted from 7:25 a.m. to 8:30 a.m. More striking than the forty-three minutes of extra sleep that these teens reported getting was the change in academic performance, indexed using a standardized measure called the Scholastic Assessment Test, or SAT.

In the year before this time change, the average verbal SAT scores of the top-performing students was a very respectable 605. The following year, after switching to an 8:30 a.m. start time, that score rose to an average 761 for the same top-tier bracket of students. Math SAT scores also improved, increasing from an average of 683 in the year prior to the time change, to 739 in the year after. Add this all up, and you see that investing in delaying school start times—allowing students more sleep and better alignment with their unchangeable biological rhythms—returned a net SAT profit of 212 points. That improvement will change which tier of university those teenagers go to, potentially altering their subsequent life trajectories as a consequence.

While some have contested how accurate or sound the Edina test case is, well-controlled and far larger systematic studies have proved that Edina is no fluke. Numerous counties in several US states have shifted the start of schools to a later hour and their students experienced significantly higher grade point averages. Unsurprisingly, performance improvements were observed regardless of time of day; however, the most dramatic surges occurred in morning classes.

It is clear that a tired, under-slept brain is little more than a leaky memory sieve, in no state to receive, absorb, or efficiently retain an education. To persist in this way is to handicap our children with partial amnesia. Forcing youthful brains to become early birds will guarantee that they do not catch the worm, if the worm in question is knowledge or good grades. We are, therefore, creating a generation of disadvantaged children, hamstrung by a privation of sleep. Later school start times are clearly, and literally, the smart choice.

One of the most troubling trends emerging in this area of sleep and brain development concerns low-income families—a trend that has direct relevance to education. Children from lower socioeconomic backgrounds are less likely to be taken to school in a car, in part because their parents often have jobs in the service industry demanding work start times at or before six a.m. Such children therefore rely on school buses for transit, and must wake up earlier than those taken to school by their parents. As a result, those already disadvantaged children become even more so because they routinely obtain less sleep than children from more affluent families. The upshot is a vicious cycle that perpetuates from one generation to the next—a closed-loop system that is very difficult to break out of. We desperately need active intervention methods to shatter this cycle, and soon.

Research findings have also revealed that increasing sleep by way of delayed school start times wonderfully increases class attendance, reduces behavioral and psychological problems, and decreases substance and alcohol use. In addition, later start times beneficially mean a later *finish* time. This protects many teens from the well-researched "danger window" between three and six p.m., when schools finish but before parents return home. This unsupervised, vulnerable time period is a recognized cause of involvement in crime and alcohol and substance abuse. Later school start times profitably shorten this danger window, reduce these adverse outcomes, and therefore lower the associated financial cost to society (a savings that could be reinvested to offset any additional expenditures that later school start times require).

Yet something even more profound has happened in this ongoing story of later school start times—something that researchers did not anticipate: the life expectancy of students increased. The leading cause of death among teenagers is road traffic accidents,* and in this regard, even the slightest dose of insufficient sleep can have marked consequences, as we have discussed. When the Mahtomedi School District of Minnesota pushed their school start time from 7:30 to 8:00 a.m., there was a 60 percent reduction in traffic accidents in drivers sixteen

*Centers for Disease Control and Prevention, "Teen Drivers: Get the Facts," Injury Prevention & Control: Motor Vehicle Safety, accessed at http://www.cdc.gov/motorvehiclesafety/teen_drivers/teendrivers_factsheet.html.

to eighteen years of age. Teton County in Wyoming enacted an even more dramatic change in school start time, shifting from a 7:35 a.m. bell to a far more biologically reasonable one of 8:55 a.m. The result was astonishing—a 70 percent reduction in traffic accidents in sixteen- to eighteen-year-old drivers.

To place that in context, the advent of anti-lock brake technology (ABS)—which prevents the wheels of a car from seizing up under hard braking, allowing the driver to still maneuver the vehicle—reduced accident rates by around 20 to 25 percent. It was deemed a revolution. Here is a simple biological factor—sufficient sleep—that will drop accident rates by more than double that amount in our teens.

These publicly available findings should have swept the education system in an uncompromising revision of school start times. Instead, they have largely been swept under the rug. Despite public appeals from the American Academy of Pediatrics and the Centers for Disease Control and Prevention, change has been slow and hard-fought. It is not enough.

School bus schedules and bus unions are a major roadblock thwarting appropriately later school start times, as is the established routine of getting the kids out the door early in the morning so that parents can start work early. These are good reasons for why shifting to a national model of later school start times is difficult. They are real pragmatic challenges that I truly appreciate, and sympathize with. But I don't feel they are sufficient excuses for why an antiquated and damaging model should remain in place when the data are so clearly unfavorable. If the goal of education is to educate, and not risk lives in the process, then we are failing our children in the most spectacular manner with the current model of early school start times.

Without change, we will simply perpetuate a vicious cycle wherein each generation of our children are stumbling through the education system in a half-comatose state, chronically sleep-deprived for years on end, stunted in their mental and physical growth as a consequence, and failing to maximize their true success potential, only to inflict that same assault on their own children decades later. This harmful spiral is only getting worse. Data aggregated over the past century from more than 650,000 schoolchildren aged five to eighteen reveal that they are sleeping two hours fewer per night than their counterparts were a hundred years ago. This is true no matter what age group, or sub-age group, you consider.

An added reason for making sleep a top priority in the education and lives of our children concerns the link between sleep deficiency and the epidemic of ADHD (attention deficit hyperactivity disorder). Children with this diagnosis are irritable, moodier, more distractible and unfocused in learning during the day, and have a significantly increased prevalence of depression and suicidal ideation. If you make a composite of these symptoms (unable to maintain focus and attention, deficient learning, behaviorally difficult, with mental health instability), and then strip away the label of ADHD, these symptoms are strongly overlapping with those caused by a lack of sleep. Take an under-slept child to a doctor and describe these symptoms without mentioning the lack of sleep, which is not uncommon, and what would you imagine the doctor is diagnosing the child with, and medicating them for? Not deficient sleep, but ADHD.

There is more irony here than meets the eye. Most people know the name of the common ADHD medications: Adderall and Ritalin. But few know what these drugs actually are. Adderall is amphetamine with certain salts mixed in, and Ritalin is a similar stimulant, called methylphenidate. Amphetamine and methylphenidate are two of the most powerful drugs we know of to prevent sleep and keep the brain of an adult (or a child, in this case) wide awake. That is the very last thing that such a child needs. As my colleague in the field, Dr. Charles Czeisler, has noted, there are people sitting in prison cells, and have been for decades, because they were caught selling amphetamines to minors on the street. However, we seem to have no problem at all in allowing pharmaceutical companies to broadcast prime-time commercials highlighting ADHD and promoting the sale of amphetamine-based drugs (e.g., Adderall, Ritalin). To a cynic, this seems like little more than an uptown version of a downtown drug pusher.

I am in no way contesting the disorder of ADHD, and not every child with ADHD has poor sleep. But we know that there are children who are sleep-deprived or suffering from an undiagnosed sleep disorder that masquerades as ADHD. They are being dosed for years of their critical development with amphetamine-based drugs.

One example of an undiagnosed sleep disorder is pediatric sleep-disordered breathing, or child obstructive sleep apnea, which is associated with heavy snoring. Overly large adenoids and tonsils can block the airway passage of a child as their breathing muscles relax during sleep. The labored snoring is the sound of turbulent air trying to be sucked down into the lungs through a semi-collapsed, fluttering airway. The resulting oxygen debt will reflexively force the brain to awaken the child briefly throughout the night so that several full breaths can be obtained, restoring full blood oxygen saturation. However, this prevents the child from reaching and/or sustaining long periods of valuable deep NREM sleep. Their sleep-disordered breathing will impose a state of chronic sleep deprivation, night after night, for months or years on end.

As the state of chronic sleep deprivation builds over time, the child will look ever more ADHD-like in temperament, cognitively, emotionally, and academically. Those children who are fortunate to have the sleep disorder recognized, and who have their tonsils removed, more often than not prove that they do not have ADHD. In the weeks after the operation, a child's sleep recovers, and with it, normative psychological and mental functioning in the months ahead. Based on recent surveys and clinical evaluations, we estimate that more than 50 percent of all children with an ADHD diagnosis actually have a sleep disorder, yet a small fraction know of their sleep condition and its ramifications. A major public health awareness campaign by governments is needed on this issue.

Stepping back from the issue of ADHD, the bigger-picture problem is ever clearer. Failed by the lack of any governmental guidelines and poor communication by researchers such as myself regarding the extant scientific data, many parents remain oblivious to the state of childhood sleep deprivation, so often undervaluing this biological necessity. A recent poll by the National Sleep Foundation affirms this point, with well over 70 percent of parents believing their child gets enough sleep, when in reality, less than 25 percent of children aged eleven to eighteen actually obtain the necessary amount.

As parents, we therefore have a jaundiced view of the need and importance of sleep in our children, sometimes even chastising or stigmatizing their desire to sleep enough, including their desperate weekend attempts to repay a sleep debt that thc school system has saddled them with through no fault of their own. I hope we can change. I hope we can break the parent-to-child transmission of sleep neglect and remove what the exhausted, fatigued brains our youth are so painfully starved of. When sleep is abundant, minds flourish. When it is deficient, they don't.

SLEEP AND HEALTH CARE

If you are about to receive medical treatment at a hospital, you'd be well advised to ask the doctor: "How much sleep have you had in the past twenty-four hours?" The doctor's response will determine, to a statistically provable degree, whether the treatment you receive will result in a serious medical error, or even death.

All of us know that nurses and doctors work long, consecutive hours, and none more so than doctors during their resident training years. Few people, however, know why. Why did we ever force doctors to learn their profession in this exhausting, sleepless way? The answer originates with the esteemed physician William Stewart Halsted, MD, who was also a helpless drug addict.

Halsted founded the surgical training program at Johns Hopkins Hospital in Baltimore, Maryland, in May 1889. As chief of the Department of Surgery, his influence was considerable, and his beliefs about how young doctors must apply themselves to medicine, formidable. There was to be a six-year residency, quite literally. The term "residency" came from Halsted's belief that doctors must live in the hospital for much of their training, allowing them to be truly committed in their learning of surgical skills and medical knowledge. Fledgling residents had to suffer long, consecutive work shifts, day and night. To Halsted, sleep was a dispensable luxury that detracted from the ability to work and learn. Halsted's mentality was difficult to argue with, since he himself practiced what he preached, being renowned for a seemingly superhuman ability to stay awake for apparently days on end without any fatigue.

But Halsted had a secret that only came to light years after his death, and helped explain both the maniacal structure of his residency program and his ability to forgo sleep. Halsted was a cocaine addict. It was a sad and apparently accidental habit, one that started years before his arrival at Johns Hopkins.

Early in his career, Halsted was conducting research on the nerve-blocking abilities of drugs that could be used as anesthetics to dull pain in surgical procedures. One of those drugs was cocaine, which prevents electrical impulse waves from shooting down the length of the nerves in the body, including those that transmit pain. Addicts of the drug know this all too well, as their nose, and often their entire face, will become numb after snorting several lines of the substance, almost like having been injected with too much anesthetic by an overly enthusiastic dentist.

Working with cocaine in the laboratory, it didn't take long before Halsted was experimenting on himself, after which the drug gripped him in a ceaseless addiction. If you read Halsted's academic report of his research findings in the *New York Medical Journal* from September 12, 1885, you'd be hard pressed to comprehend it. Several medical historians have suggested that the writing is so discombobulated and frenetic that he undoubtedly wrote the piece when high on cocaine.

Colleagues noticed Halsted's odd and disturbing behaviors in the years before and after his arrival at Johns Hopkins. This included excusing himself from the operating theater while he was supervising residents during surgical procedures, leaving the young doctors to complete the operation on their own. At other times, Halsted was not able to operate himself because his hands were shaking so much, the cause of which he tried to pass off as a cigarette addiction.

Halsted was now in dire need of help. Ashamed and nervous that his colleagues would discover the truth, he entered a rehabilitation clinic under his first and middle name, rather than using his surname. It was the first of many unsuccessful attempts at kicking his habit. For one stay at Butler Psychiatric Hospital in Providence, Rhode Island, Halsted was given a rehabilitation program of exercise, a healthy diet, fresh air, and, to help with the pain and discomfort of cocaine withdrawal, morphine. Halsted subsequently emerged from the "rehabilitation" program with both a cocaine addiction *and* a morphine addiction. There were even stories that Halsted would inexplicably send his shirts to be laundered in Paris, and they would return in a parcel containing more than just pure-white shirts.

Halsted inserted his cocaine-infused wakefulness into the heart of Johns Hopkins's surgical program, imposing a similarly unrealistic mentality of sleeplessness upon his residents for the duration of their training. The exhausting residency program, which persists in one form or another throughout all US medical schools to this day, has left countless patients hurt or dead in its wake—and likely residents, too. That may sound like an unfair charge to level considering the wonderful, lifesaving work our committed and caring young doctors and medical staff perform, but it is a provable one.

Many medical schools used to require residents to work thirty hours. You may think that's short, since I'm sure you work at least forty hours a week. But for residents, that was thirty hours all in one go. Worse, they often had to do two of these thirty-hour continuous shifts within a week, combined with several twelve-hour shifts scattered in between.

The injurious consequences are well documented. Residents working a thirty-hour-straight shift will commit 36 percent more serious medical errors, such as prescribing the wrong dose of a drug or leaving a surgical implement inside of a patient, compared with those working sixteen hours or less. Additionally, after a thirty-hour shift without sleep, residents make 460 percent more diagnostic mistakes in the intensive care unit than when well rested after enough sleep. Throughout the course of their residency, one in five medical residents will make a sleepless-related medical error that causes significant, liable harm to a patient. One in twenty residents will kill a patient due to a lack of sleep. Since there are over 100,000 residents currently in training in US medical programs, this means that many hundreds of people—sons, daughters, husbands, wives, grandparents, brothers, sisters—are needlessly losing their lives every year because residents are not allowed to get the sleep they need. As I write this chapter, a new report has discovered that medical errors are the third-leading cause of death among Americans after heart attacks and cancer. Sleeplessness undoubtedly plays a role in those lives lost.

Young doctors themselves can become part of the mortality statistics. After a thirty-hour continuous shift, exhausted residents are 73 percent more likely to stab themselves with a hypodermic needle or cut themselves with a scalpel, risking a blood-born infectious disease, compared to their careful actions when adequately rested.

One of the most ironic statistics concerns drowsy driving. When a sleep-deprived resident finishes a long shift, such as a stint in the ER trying to save victims of car accidents, and then gets into their own car to drive home, their chances of being involved in a motor vehicle accident are increased by 168 percent because of fatigue. As a result, they may find themselves back in the very same hospital and ER from which they departed, but now as a victim of a car crash caused by a microsleep.

Senior medical professors and attending physicians suffer the same bankruptcy of their medical skills following too little sleep. For example, if you are a patient under the knife of an attending physician who has not been allowed at least a six-hour sleep opportunity the night prior, there is a 170 percent increased risk of that surgeon inflicting a serious surgical error on you, such as organ damage or major hemorrhaging, relative to the superior procedure they would conduct when they have slept adequately.

If you are about to undergo an elective surgery, you should ask how much sleep your doctor has had and, if it is not to your liking, you may not want to proceed. No amount of years on the job helps a doctor "learn" how to overcome a lack of sleep and develop resilience. How could it? Mother Nature spent millions of years implementing this essential physiological need. To think that bravado, willpower, or a few decades of experience can absolve you (a surgeon) of an evolutionarily ancient necessity is the type of hubris that, as we know from the evidence, costs lives.

The next time you see a doctor in a hospital, keep in mind the study we have previously discussed, showing that after twenty-two hours without sleep, human performance is impaired to the same level as that of someone who is legally drunk.

Why haven't these, and now many similar such findings, triggered a responsible revision of work schedules for residents and attending physicians by the American medical establishment? Why are we not giving back sleep to our exhausted and thus error-prone doctors? The collective goal is, after all, to achieve the highest quality of medical practice and care, is it not?

Facing government threats that would apply federally enforced work hours due to the extent of damning evidence, the Accreditation Council for Graduate Medical Education made the following alterations. First-year residents would be limited to (1) working no more than an 80-hour week (which still averages out at 11.5 hours per day for 7 days straight), (2) working no more than 24 hours nonstop, and (3) performing one overnight on-call shift every third night. That revised schedule still far exceeds any ability of the brain to perform optimally. Errors, mistakes, and deaths continued in response to the anemic diet of sleep they were being fed while training. As the research studies kept accumulating, the Institute of Medicine, part of the US National Academy of Sciences, issued a report with a clear statement: working for more than sixteen consecutive hours without sleep is hazardous for both the patient and resident physician.

You may have noticed my specific wording in the above paragraph: *first*-year residents. This is because the revised rule (at the time of writing this book) has only been applied to those in their first year of training, and not to those in later years of a medical residency. Why? Because the Accreditation Council for Graduate Medical Education—the elite board of high-powered physicians that dictates the American residency training structure—stated that data proving the dangers of insufficient sleep had only been gathered in residents in their first year of the program. As a result, they felt there was no evidence to justify a change for residents in years two to five—as if getting past the twelve-month point in a medical residency program magically confers immunity against the biological and psychological effects of sleep deprivation—effects that these same individuals had previously been so provably vulnerable to just months before.

Of course, medical institutions put forward understandable arguments to justify the old-school way of sleep abuse. The most common harkens back to a William Halsted–like mind-set: without working exhaustive shifts, it will take far too long to train residents, and they will not learn as effectively. Why, then, can several western European countries train their young doctors within the same time frame when they are limited to working no more than forty-eight hours in one week, without continuous long periods of sleeplessness? Perhaps they are just not as well trained? This, too, is erroneous, since many of those western European medical programs, such as in the UK and Sweden, rank among the top ten countries for most medical practice health outcomes, while the majority of US institutes rank somewhere between eighteenth and thirty-second. As a matter of fact, several pilot studies in the US have shown that when you limit residents to no more than a sixteen-hour shift, with at least an eight-hour rest opportunity before the next shift,* the number of serious medical errors made—defined as causing or having the potential to cause harm to a patient—drops by over 20 percent. Furthermore, residents made significantly fewer diagnostic errors to begin with.

There does not seem to be a particularly strong evidence-based argument for persisting with the current sleep-anemic model of medical training, one that debilitates the learning, health, and safety of young doctors and patients alike. That it remains this way in the stoic grip of senior medical officials could perhaps be a case of "my mind is made up, don't confuse me with the facts."

More generally, I feel we as a society can work toward dismantling our negative and counterproductive attitude toward sleep: one that is epitomized in the words of a US senator who once said, "I've always loathed the necessity of sleep. Like death, it puts even the most powerful men on their backs." This attitude encapsulates some of the modern views of sleep: loathsome, annoying, enfeebling. Though the senator in question is a television character called Frank Underwood from the series

*Based on this description, you could be forgiven for thinking that residents now have a delightful eight-hour sleep opportunity. Unfortunately, this is not true. During that eight-hour break, residents are supposed to return home, eat, spend time with significant others, perform any physical exercise they desire, sleep, shower, and commute back to the hospital. It's hard to imagine getting much more than five hours of shut-eye amid all that must happen in between—which, indeed, they don't. A maximum twelve-hour shift, with a twelve-hour break, is the very most we should be asking of a resident, or any attending doctor, for that matter.

House of Cards, the writers have—biographically, I believe—placed their fingers on the very nub of the sleep-neglect problem.

Tragically, this same neglect has resulted in some of the worst global catastrophes punctuating the human historical record. Consider the infamous reactor meltdown at the Chernobyl nuclear power station on April 26, 1986. The radiation from the disaster was one hundred times more powerful than the atomic bombs dropped in World War II. There is speculation the operators were sleep deprived, working an exhaustive shift, occurring, without coincidence, at one a.m. Thousands died from the long-term effects of radiation in the protracted decades following the event, and tens of thousands more suffered a lifetime of debilitating medical and developmental ill health. We can also recount the *Exxon Valdez* oil tanker that ran aground on Bligh Reef in Alaska on March 24, 1989, breaching its hull. An estimated 10 million to 40 million gallons of crude oil spilled across a 1,300-mile range of the surrounding shoreline. Left dead were more than 500,000 seabirds, 5,000 otters, 300 seals, over 200 bald eagles, and 20 orca whales. The coastal ecosystem has never recovered. Early reports suggested that the captain was inebriated while navigating the vessel. Later, however, it was revealed that the sober captain had turned over command to a qualified mate on deck, who had only slept six out of the previous forty-eight hours, causing him to make the cataclysmic navigational error.

Both of these global tragedies were entirely preventable. The same is true for every sleep-loss statistic in this chapter.

CHAPTER 16

A New Vision for Sleep in the Twenty-First Century

That our lack of sleep could be considered as a slow form of self-euthanasia, what can be done about it? In this book, I have described the problems and causes of our collective sleeplessness. But what of solutions? How can we effect change?

For me, addressing this issue involves two steps of logic. First, we must understand why the problem of deficient sleep seems to be so resistant to change, and thus persists and grows worse. Second, we must develop a structured model for effecting change at every possible leverage point we can identify. There is not going to be a single, magic-bullet solution. After all, there is not just one reason for why society is collectively sleeping too little, but many. Below, I sketch out a new vision for sleep in the modern world—a road map of sorts that ascends through numerous levels of intervention opportunities, visualized in figure 17.

Figure 17: Levels of Sleep Intervention

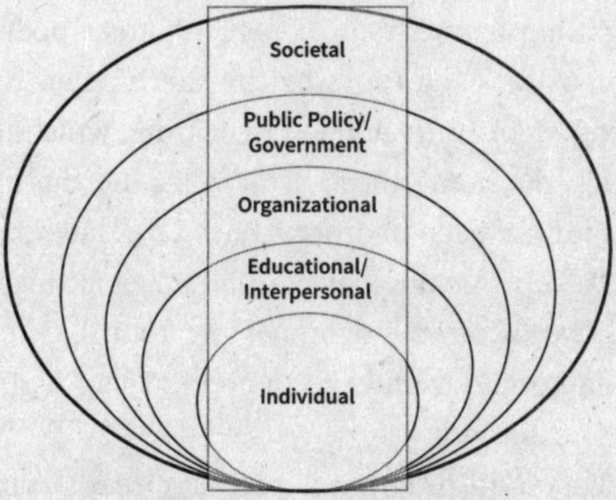

INDIVIDUAL TRANSFORMATION

Increasing sleep for an individual can be achieved through both passive methods, which require no effort from the individual and are thus preferable, and active methods, which do. Here are several possibilities that may not be so far-fetched, all of which build on proven scientific methods for enhancing sleep quantity and quality.

The intrusion of technology into our homes and bedrooms is claimed by many of my research colleagues to be robbing us of precious sleep, and I agree. Evidence discussed in this book, such as the harmful effects of LED-emitting devices at night, proves this to be true. Some have therefore lobbied to keep sleep analog, as it were, in this increasingly digital world, leaving technology out of the discussion.

Here, however, I actually disagree. Yes, the future of sleep is about a return to the past in the sense that we must reunite with regular, plentiful sleep, as we once knew a century ago. But to battle against rather than unite with technology is the wrong approach in my mind. For one thing, it's a losing battle: we will never put that technological genie back into its bottle, nor do we need to. Instead, we can use this powerful tool to our advantage. Within three to five years, I am quite certain there will be commercially available, affordable devices that track an individual's sleep and circadian rhythm with high accuracy. When that happens, we can marry these individual sleep trackers with the revolution of in-home networked devices like thermostats and lighting. Some are already trying to do this as I write.

Two exciting possibilities unfold. First, such devices could compare the sleep of each family member in each separate bedroom with the temperature sensed in each room by the thermostat. Using common machine-learning algorithms applied over time, we should be able to intelligently teach the home thermostat what the thermal sweet spot is for each occupant in each bedroom, based on the biophysiology calculated by their sleep-tracking device (perhaps splitting the difference when there are two or more individuals per room). Granted there are many different factors that make for a good or bad night of sleep, but temperature is one of them.

Better still, we could program a natural circadian lull and rise in temperature across the night that is in harmony with each body's expectations, rather than the constant nighttime temperature set in most homes and apartments. Over time, we could intelligently curate a tailored thermal sleep environment that is personalized to the circadian rhythms of each individual occupant of each bedroom, departing from the unhelpful non-varying thermal backdrop that plagues the sleep of most people using standard home thermostats. Both these changes require no effort from an individual, and should hasten the speed of sleep onset, increase total sleep time, and even deepen NREM-sleep quality for all household members (as discussed in chapter 13).

The second passive solution concerns electric light. Many of us suffer from overexposure to nighttime light, particularly blue-dominant LED light from our digital devices. This evening digital light suppresses melatonin and delays our sleep timing. What if we can turn that problem into a solution? Soon, we should be able to engineer LED bulbs with filters that can vary the wavelength of light that they emit, ranging from warm yellow colors less harmful to melatonin, to strong blue light that powerfully suppresses it.

Paired with sleep trackers that can accurately characterize our personal biological rhythms, we can install these new bulbs throughout a home, all connected to the home network. The lightbulbs (and even other networked LED-screen devices, such as iPads) would be instructed to gradually dial down the harmful blue light in the home as the evening progresses, based on an individual's (or set of individuals') natural sleep-wake pattern. We could do this dynamically and seamlessly as individuals move from one room to the next in real time. Here again we can intelligently split the difference on the fly based on the biophysiological mix of whoever is in the room. In doing so, the users' own brains and bodies, measured and translated through the wearables to the networked home, would synergistically regulate light and thus melatonin release that promotes, rather than impedes, optimal regulation of sleep for one and all. It is a vision of personalized sleep medicine.

Come the morning, we can reverse this trick. We can now saturate our indoor environments with powerful blue light that shuts off any lingering melatonin. This will help us wake up faster, more alert, and with a brighter mood, morning after morning.

We could even use this same light-manipulation idea to apply a slight nudge in someone's sleep-wake rhythm within a biologically reasonable range (plus or minus thirty to forty minutes), should they desire, gradually moving it earlier or later. For example, if you have an unusually early morning meeting in the middle of the workweek, this technology, synched to your online calendar, would gradually begin shifting you (your circadian rhythm) to a slightly earlier bed and rise time starting on Monday. This way, that early-morning rise time on Wednesday won't be as miserable, or cause such biological turmoil within your brain and body. This would be equally, if not more, applicable in helping individuals overcome jet lag when traveling between time zones, all dispensed through LED-emitting personal devices that people already travel with—phones, tablets, laptop computers.

Why stop at the home environment or in the infrequent circumstance of jet lag? Cars can adopt these same lighting solutions to help manipulate alertness during morning commutes. Some of the highest rates of drowsy-driving accidents occur during mornings, especially early mornings. What if car cockpits could be bathed in blue light during early-morning commutes? The levels would have to be tempered so as not to distract the driver or others on the road, but you'll recall from chapter 13 that one does not need especially bright light (lux) to have a measurable impact of melatonin suppression and enhanced wakefulness. This idea could be particularly helpful in those parts of the Northern and Southern Hemispheres during their respective winter mornings where this issue is most problematic. In the workplace, for those lucky enough to have their own office, lighting rhythm could be custom fit to the occupant using the same principles. But even cubicles, which are not so different from the cell of a car, could be personally tailored in this light-dependent manner, based on the individual sitting in that cubicle.

How much benefit such changes would make remains to be proven, but I can already tell you of some data from ever-sleep-sensitive NASA, with which I worked on sleep issues early in my career. Astronauts on the International Space Station travel through space at 17,500 miles per hour and complete an orbit of the Earth once every ninety to one hundred minutes. As a result, they experience "daylight" for about fifty minutes, and "night" for about fifty minutes. Although astronauts are therefore treated to the delight of a sunrise and sunset sixteen times a day, it wreaks utter havoc on their sleep-wake rhythms, causing terrible issues with insomnia and sleepiness. Make a mistake at your job on planet Earth, and your boss may reprimand you. Make a mistake in a long metal tube floating through the vacuum of space with payloads and mission costs in the hundreds of millions, and the consequences can be much, much worse.

To combat this issue, NASA began collaborating with a large electrical company some years ago to create just the types of special lightbulbs I describe. The bulbs were to be installed in the space station to bathe the astronauts in a much more Earth-like cycle of twenty-four-hour light and dark. With regulated environmental light came a superior regulation of the astronauts' biological melatonin rhythms, including their sleep, thereby reducing operations errors associated with fatigue. I must admit that the development cost of each lightbulb was in the neighborhood of $300,000. But numerous companies are now hard at work constructing similar bulbs for a fraction of that cost. The first iterations are just starting to come to market as I write. When costs become more competitive with standard bulbs, these and many other possibilities will become a reality.

Solutions that are less passive, requiring an individual to actively participate in change, will be harder to institute. Human habits, once established, are difficult to change. Consider the countless New Year's resolutions you've made but never kept. Promises to stop the overeating, to get regular exercise, or to quit smoking are but a few examples of habits we often want to change to prevent ill health, yet rarely succeed at actually changing. Our persistence in sleeping too little may similarly appear to be a lost cause, but I am optimistic that several active solutions will make a real difference for sleep.

Educating people about sleep—through books, engaging lectures, or television programs—can help combat our sleep deficit. I know firsthand from teaching a class on the science of sleep to four hundred to five hundred undergraduates each semester. My students complete an anonymous sleep survey at the start and the end of the course. Across a semester of lectures, the amount of sleep they report getting increases by forty-two minutes per night on average. Trivial as that may sound, it does translate to five hours of extra sleep each week, or seventy-five extra hours of sleep each semester.

But this isn't enough. I'm sure a depressingly large proportion of my students returned to their shorter, unhealthy sleep habits in the years after. Just as describing the scientific dangers of how eating junk food leads to obesity rarely ends up with people choosing broccoli over a cookie, knowledge alone is not enough. Additional methods are required.

One practice known to convert a healthy new habit into a permanent way of life is exposure to your own data. Research in cardiovascular disease is a good example. If patients are given tools that can be used at home to track their improving physiological health in response to an exercise plan—such as blood pressure monitors during exercise programs, scales that log body mass index during dieting efforts, or spirometry devices that register respiratory lung capacity during attempted smoking cessation—compliance rates with rehabilitation programs increase. Follow up with those same individuals after a year or even five, and more of them have maintained their positive change in lifestyle and behavior as a consequence. When it comes to the quantified self, it's the old adage of "seeing is believing" that ensures longer-term adherence to healthy habits.

With wearables that accurately track our slumber fast emerging, we can apply this same approach to sleep. Harnessing smartphones as a central hub to gather an individual's health data from various sources—physical activity (such as number of steps or minutes and intensity of exercise), light exposure, temperature, heart rate, body weight, food intake, work productivity, or mood—we show each individual how their own sleep is a direct predictor of their own physical and mental health. It's likely that, if you wore such a device, you would find out that on the nights you slept more you ate less food the next day, and of a healthy kind; felt brighter, happier, and more positive; had better relationship interactions; and accomplished more in less time at work. Moreover, you would discover that during months of the year when you were averaging more sleep, you were sick less; your weight, blood pressure, and medication use were all lower; and your relationship or marriage satisfaction, as well as sex life, were better.

Reinforced day after day, month after month, and ultimately year after year, this nudge could change many people's sleep neglect for the better. I'm not so naïve to think it would be a radical change, but if this increased your sleep amount by just fifteen to twenty minutes each night, the science indicates that it would make a significant difference across the life span and save trillions of dollars within the global economy at the population level, to name but two benefits. It could be one of the most powerful factors in a future vision that shifts from a model of sick care (treatment), which is what we do now, to health care (prevention)—the latter aiming to stave off a need for the former. Prevention is far more efficient than treatment, and costs far less in the long run.

Going even further, what if we moved from a stance of *analytics* (i.e., here is your past and/or current sleep and here is your past and/or current body weight) to that of forward-looking *predictalytics*? To explain the term, let me go back to the smoking example. There are efforts to create predictalytics apps that start with you taking a picture of your own face with the camera of your smartphone. The app then asks you how many cigarettes you smoke on average a day. Based on scientific data that understand how smoking quantity impacts outward health features such as bags under your eyes, wrinkles, psoriasis, thinning hair, and yellowed teeth, the app predictively modifies your face on the assumption of your continued smoking, and does so at different future time points: one year, two years, five years, ten years.

The very same approach could be adopted for sleep, but at many different levels: outward appearance as well as inward brain and body health. For example, we could show individuals their increasing risk (albeit non-deterministic) of conditions such as Alzheimer's disease or certain cancers if they continue sleeping too little. Men could see projections on how much their testosterone level will drop should their sleep neglect continue. Similar risk predictions could be made for gains in body weight, diabetes, or immune impairment and infection.

Another example involves offering individuals a prediction of when they should or should not get their flu shot based on sleep amount in the week prior. You will recall from chapter 8 that getting four to six hours of sleep a night in the week before your flu shot means that you will produce less than half of the normal antibody response required, while seven or more hours of sleep consistently returns a powerful and comprehensive immunization response. The goal would be to unite health-care providers and hospitals with real-time updates on an individual's sleep, week to week. Through notifications, the software will identify the optimal time for when an individual should get their flu shot to maximize vaccination success.

Not only will this markedly improve an individual's immunity but also that of the community, through developing more effective "herd immune benefits." Few people realize that the annual financial cost of the flu in the US is around $100 billion ($10 billion direct and $90 billion in lost work productivity). Even if this software solution decreases flu infection rates by just a small percentage, it will save hundreds of millions of dollars by way of improved immunization efficiency by reducing the cost burden on hospital services, both the inpatient and outpatient service utilization. By avoiding lost productivity through illness and absenteeism during the flu season, businesses and the economy stand to save even more—potentially billions of dollars—and could help subsidize the effort.

We can scale this solution globally: anywhere there is immunization and the opportunity to track an individual's sleep, there is the chance for marked cost savings to health-care systems, governments, and businesses, all with the motivated goal of trying to help people live healthier lives.

EDUCATIONAL CHANGE

Over the past five weeks, I conducted an informal survey of colleagues, friends, and family in the United States and in my home country of the United Kingdom. I also sampled friends and colleagues from Spain, Greece, Australia, Germany, Israel, Japan, South Korea, and Canada.

I asked about the type of health and wellness education they received at school when they were growing up. Did they receive instruction on diet? Ninety-eight percent of them did, and many still remembered some details (even if those are changing based on current recommendations). Did they receive tutelage on drugs, alcohol, safe sex, and reproductive health? Eighty-seven percent said yes. Was the importance of exercise impressed upon them at some point during their schooling, and/or was the practice of physical education activities mandatory on a weekly basis? Yes—100 percent of people confirmed it was.

This is hardly a scientific data set, but still, some form of dietary, exercise, and health-related schooling appears to be part of a worldwide educational plan that most children in developed nations receive.

When I asked this same diverse set of individuals if they had received any education about sleep, the response was equally universal in the opposite direction: 0 percent received any educational materials or information about sleep. Even in the health and personal wellness education that some individuals described, there was nothing resembling lip service to sleep's physical or mental health importance. If these individuals are representative, it suggests that sleep holds no place in the education of our children. Generation after generation, our young minds continue to remain unaware of the immediate dangers and protracted health impacts of insufficient sleep, and I for one feel that is wrong.

I would be keen to work with the World Health Organization to develop a simple educational module that can be implemented in schools around the world. It could take many forms, based on age group: an animated short accessible online, a board game in physical or digital form (one that could even be played internationally with sleep "pen pals"), or a virtual environment that helps you explore the secrets of sleep. There are many options, all of them easily translatable across nations and cultures.

The goal would be twofold: change the lives of those children and, by way of raising sleep awareness and better sleep practice, have that child pass on their healthy sleep values to their own children. In this way, we would begin a familial transmission of sleep appreciation from one generation to the next, as we do with things like good manners and morality. Medically, our future generations would not only enjoy a longer life span, but, more importantly, a longer health span, absolved of the mid- and late-life diseases and disorders that we know are caused by (and not simply associated with) chronic short sleep. The cost of delivering such sleep education programs would be a tiny fraction of what we currently pay for our unaddressed global sleep deficit.

ORGANIZATIONAL CHANGE

Let me offer three rather different examples for how we could achieve sleep reform in the workplace and key industries.

First, to employees in the workplace. The giant insurance company Aetna, which has almost fifty thousand employees, has instituted the option of bonuses for getting more sleep, based on verified sleep-tracker data. As Aetna chairman and CEO Mark Bertolini described, "Being present in the workplace and making better decisions has a lot to do with our business fundamentals." He further noted, "You can't be prepared if you're half asleep." If workers string together twenty seven-hour nights of sleep or more in a row, they receive a twenty-five-dollar-per-night bonus, for a (capped) total of five hundred dollars.

Some may scoff at Bertolini's incentive system, but developing a new business culture that takes care of the entire life cycle of an employee, night and day, is as economically prudent as it is compassionate. Bertolini seems to know that the net company benefit of a well-slept employee is considerable. The return on the sleep investment in terms of productivity, creativity, work enthusiasm, energy, efficiency—not to mention happiness, leading to people wanting to work at your institution, and stay—is undeniable. Bertolini's empirically justified wisdom overrides misconceptions about grinding down employees with sixteen- to eighteen-hour workdays, burning them out in a model of disposability and declining productivity, littered with sick days, all the while triggering low morale and high turnover rates.

I wholeheartedly endorse Bertolini's idea, though I would modify it in the following way. Rather than—or as an alternative to—providing financial bonuses, we could offer added vacation time. Many individuals value time off more than modest financial perks. I would suggest a "sleep credit system," with sleep time being exchanged for either financial bonuses or extra vacation days. There would be at least one proviso: the sleep credit system would not simply be calculated on total hours clocked during one week or one month. As we have learned, sleep *continuity*—consistently getting seven to nine hours of sleep opportunity each night, every night, without running a debt during the week and hoping to pay it off by binge-sleeping at the weekend—is just as important as total sleep time if you are to receive the mental and physical health benefits of sleep. Thus, your "sleep credit score" would be calculated based on a combination of sleep *amount* and night-to-night sleep *continuity*.

Those with insomnia need not be penalized. Rather, this method of routine sleep tracking would help them identify this issue, and cognitive behavioral therapy could be provided through their smartphones. Insomnia treatment could be incentivized with the same credit benefits, further improving individual health and productivity, creativity, and business success.

The second change-idea concerns flexible work shifts. Rather than required hours with relatively hard boundaries (i.e., the classic nine to five), businesses need to adapt a far more tapered vision of hours of operation, one that resembles a squished inverted-U shape. Everyone would be present during a core window for key interactions—say, twelve to three p.m. Yet there would be flexible tail ends either side to accommodate all individual chronotypes. Owls could start work late (e.g., noon) and continue into the evening, giving their full force of mental capacity and physical energy to their jobs. Larks can likewise do so with early start and finish times, preventing them from having to coast through the final hours of the "standard" workday with inefficient sleepiness. There are secondary benefits. Take rush-hour traffic as just one example, which would be lessened in both the morning and evening phases. The indirect cost savings of time, money, and stress would not be trivial.

Maybe your workplace claims to offer some version of this. However, in my consulting experience, the opportunity might be suggested but is rarely embraced as acceptable, especially in the eyes of managers and leaders. Dogmas and mind-sets appear to be one of the greatest rate-limiting barriers preventing better (i.e., sleep-smart) business practices.

The third idea for sleep change within industry concerns medicine. As urgent as the need to inject more sleep in residents' work schedules is the need to radically rethink how sleep factors into patient care. I can illuminate this idea with two concrete examples.

EXAMPLE 1—PAIN

The less sleep you have had, or the more fragmented your sleep, the more sensitive you are to pain of all kinds. The most common place where people experience significant and sustained pain is often the very last place they can find sound sleep: a hospital. If you have been unfortunate enough to spend even a single night in the hospital, you will know this all too well. The problems are especially compounded in the intensive care unit, where the most severely sick (i.e., those most in need of sleep's help) are cared for. Incessant beeping and buzzing from equipment, sporadic alarms, and frequent tests prevent anything resembling restful or plentiful sleep for the patient.

Occupational health studies of inpatient rooms and wards report a decibel level of sound pollution that is equivalent to that of a noisy restaurant or bar, twenty-four hours a day. As it turns out, 50 to 80 percent of all intensive care alarms may be unnecessary or ignorable by staff. Additionally frustrating is that not all tests and patient checkups are time sensitive, yet many are ill-timed with regard to sleep. They occur either during afternoon times when patients would otherwise be enjoying a natural, biphasic-sleep nap, or during early-morning hours when patients are only now settling into solid sleep.

Little surprise that across cardiac, medical, and surgical intensive care units, studies consistently demonstrate uniformly bad sleep in all patients. Upset by the noisy, unfamiliar ICU environment, sleep takes longer to initiate, is littered with awakenings, is shallower in depth, and contains less overall REM sleep. Worse still, doctors and nurses consistently overestimate the amount of sleep they think patients obtain in intensive care units, relative to objectively measured sleep in these individuals. All told, the sleep environment, and thus sleep amount, of a patient in this hospital environment is entirely antithetical to their convalescence.

We can solve this. It should be possible to design a system of medical care that places sleep at the center of patient care, or close to it. In one of my own research studies, we have discovered that pain-related centers within the human brain are 42 percent more sensitive to unpleasant thermal stimulation (non-damaging, of course) following a night of sleep deprivation, relative to a full, healthy eight-hour night of sleep. It is interesting to note that these pain-related brain regions are the same areas that narcotic medications, such as morphine, act upon. Sleep appears to be a natural analgesic, and without it, pain is perceived more acutely by the brain, and, most importantly, felt more powerfully by the individual. Morphine is not a desirable medication, by the way. It has serious safety issues related to the cessation of breathing, dependency, and withdrawal, together with terribly unpleasant side effects. These include nausea, loss of appetite, cold sweats, itchy skin, and urinary and bowel issues, not to mention a form of sedation that prevents natural sleep. Morphine also alters the action of other medications, resulting in problematic interaction effects.

Extrapolating from a now extensive set of scientific research, we should be able to reduce the dose of narcotic drugs on our hospital wards by improving sleep conditions. In turn, this would lessen safety risks, reduce the severity of side effects, and decrease the potential for drug interactions.

Improving sleep conditions for patients would not only reduce drug doses, it would also boost their immune system. Inpatients could therefore mount a far more effective battle against infection and accelerate postoperative wound healing. With hastened recovery rates would come shorter inpatient stays, reducing health-care costs and health insurance rates. Nobody wants to be in the hospital any longer than is absolutely necessary. Hospital administrators feel likewise. Sleep can help.

The sleep solutions need not be complicated. Some are simple and inexpensive, and the benefits should be immediate. We can start by removing any equipment and alarms that are not necessary for any one patient. Next, we must educate doctors, nurses, and hospital administrations on the scientific health benefits of sound sleep, helping them realize the premium we must place on patients' slumber. We can also ask patients about their regular sleep schedules on the standard hospital admission form, and then structure assessments and tests around their habitual sleep-wake rhythms as much as possible. When I'm recovering from an appendicitis operation, I certainly don't want to be woken up at 6:30 a.m. when my natural rise time is 7:45 a.m.

Other simple practices? Supply all patients with earplugs and a face mask when they first come onto a ward, just like the complimentary air travel bag you are given on long-haul flights. Use dim, non-LED lighting at night and bright lighting during the day. This will help maintain strong circadian rhythms in patients, and thus a strong sleep-wake pattern. None of these is especially costly; most of them could happen tomorrow, all of them to the significant benefit of a patient's sleep, I'm certain.

EXAMPLE 2—NEONATES

To keep a preterm baby alive and healthy is a perilous challenge. Instability of body temperature, respiratory stress, weight loss, and high rates of infection can lead to cardiac instability, neurodevelopment impairments, and death. At this premature stage of life, infants should be sleeping the vast majority of the time, both day and night. However, in most neonatal intensive care units, strong lighting will often remain on throughout the night, while harsh electric overhead light assaults the thin eyelids of these infants during the day. Imagine trying to sleep in constant light for twenty-four hours a day. Unsurprisingly, infants do not sleep normally under these conditions. It is worth reiterating that which we learned in the chapter on the effects of sleep deprivation in humans and rats: a loss in the ability to maintain core body temperature, cardiovascular stress, respiratory suppression, and a collapse of the immune system.

Why are we not designing NICUs and their care systems to foster the very highest sleep amounts, thereby using sleep as the lifesaving tool that Mother Nature has perfected it to be? In just the last few months, we have preliminary research findings from several NICUs that have implemented dim-lighting conditions during the day and near-blackout conditions at night. Under these conditions, infant sleep stability, time, and quality all improved. Consequentially, marked and significant improvements in neonate weight gain and significantly higher oxygen saturation levels in blood were observed, relative to those preterms who did not have their sleep prioritized and thus regularized. Better still, these well-slept preterm babies were also discharged from the hospital several weeks earlier!

We can also implement this strategy in underdeveloped countries without the need for costly lighting changes by simply placing a darkening piece of plastic—a light-diffusing shroud, if you will—over neonatal cots. The cost is less than $1, but will have a significant, lux-reducing benefit, stabilizing and enhancing sleep. Even something as simple as bathing a young child at the right time before bed (rather than in the middle of the night, as I've seen occur) would help foster, rather than perturb, good sleep. Both are globally viable methods.

I must add that there is nothing stopping us from prioritizing sleep in similarly powerful ways across all pediatric units for all children in all countries.

PUBLIC POLICY AND SOCIETAL CHANGE

At the highest levels, we need better public campaigns educating the population about sleep. We spend a tiny fraction of our transportation safety budget warning people of the dangers of drowsy driving compared with the countless campaigns and awareness efforts regarding accidents linked to drugs or alcohol. This despite the fact that drowsy driving is responsible for more accidents than either of these two issues—and is more deadly. Governments could save hundreds of thousands of lives each year if they mobilized such a campaign. It would easily pay for itself, based on the cost savings to the health-care and emergency services bills that drowsy-driving accidents impose. It would of course help lower health-care and auto insurance rates and premiums for individuals.

Prosecutorial law regarding drowsy driving is another opportunity. Some states have a vehicular manslaughter charge associated with sleep deprivation, which is of course far harder to prove than blood alcohol level. Having worked with several large automakers, I can report that soon we will have smart technology inside of cars that may help us know, from a driver's reactions, eyes, driving behavior, and the nature of the crash, what the prototypical "signature" is of a clearly drowsy-driving accident. Combined with a personal history, especially as personal sleep-tracking devices become more popular, we may be very close to developing the equivalent of a Breathalyzer for sleep deprivation.

I know that may sound unwelcome to some of you. But it would not if you had lost a loved one to a fatigue-related accident. Fortunately, the rise of semiautonomous-driving features in cars can help us avoid this issue. Cars can use these very same signatures of fatigue to heighten their watch and, when needed, take greater self-control of the vehicle from the driver.

At the very highest levels, transforming entire societies will be neither trivial nor easy. Yet we can borrow proven methods from other areas of health to shift society's sleep for the better. I offer just one example. In the United States, many health insurance companies provide a financial credit to their members for joining a gym. Considering the health benefits of increased sleep amount, why don't we institute a similar incentive for racking up more consistent and plentiful slumber? Health insurance companies could approve valid commercial sleep-tracking devices that individuals commonly own. You, the individual, could then upload your sleep credit score to your health-care provider profile. Based on a tiered, pro-rata system, with reasonable threshold expectations for different age groups, you would be awarded a lower insurance rate with increasing sleep credit on a month-to-month basis. Like exercise, this in turn will help improve societal health en masse and lower the cost of health-care utilization, allowing people to have longer and healthier lives.

Even with lower insurance paid by the individual, health insurance companies would still gain, as it would significantly decrease the cost burden of their insured individuals, allowing for greater profit margins. Everyone wins. Of course, just like a gym membership, some people will start off adhering to the regime but then stop, and some may look for ways to bend or play the system regarding accurate sleep assessment. However, even if only 50 to 60 percent of individuals truly increase their sleep amount, it could save tens or hundreds of millions of dollars in terms of health costs—not to mention hundreds of thousands of lives.

• • •

This tour of ideas offers, I hope, some message of optimism rather than the tabloid-like doom with which we are so often assaulted in the media regarding all things health. More than hope, however, I wish for it to spark better sleep solutions of your own; ideas that some of you may translate into a non- or for-profit commercial venture, perhaps.

Conclusion

To Sleep or Not to Sleep

Within the space of a mere hundred years, human beings have abandoned their biologically mandated need for adequate sleep—one that evolution spent 3,400,000 years perfecting in service of life-support functions. As a result, the decimation of sleep throughout industrialized nations is having a clear impact on our health, our life expectancy, our safety, our productivity, and the education of our children.

This silent sleep loss epidemic is one of the new public health challenges we face in the twenty-first century in developed nations. If we wish to avoid the suffocating noose of sleep neglect, the premature death it inflicts, and the sickening health it invites, a radical shift in our personal, cultural, professional, and societal appreciation of sleep must occur.

I believe it is time for us to reclaim our right to a full night of sleep, without embarrassment or the damaging stigma of laziness. In doing so, we can be reunited with that most powerful elixir of wellness and vitality, dispensed through every conceivable biological pathway. Then we may remember what it feels like to be truly awake during the day, infused with the very deepest plenitude of being.

Appendix

*Twelve Tips for Healthy Sleep**

1. Stick to a sleep schedule. Go to bed and wake up at the same time each day. As creatures of habit, people have a hard time adjusting to changes in sleep patterns. Sleeping later on weekends won't fully make up for a lack of sleep during the week and will make it harder to wake up early on Monday morning. Set an alarm for bedtime. Often we set an alarm for when it's time to wake up but fail to do so for when it's time to go to sleep. If there is only one piece of advice you remember and take from these twelve tips, this should be it.
2. Exercise is great, but not too late in the day. Try to exercise at least thirty minutes on most days but not later than two to three hours before your bedtime.
3. Avoid caffeine and nicotine. Coffee, colas, certain teas, and chocolate contain the stimulant caffeine, and its effects can take as long as eight hours to wear off fully. Therefore, a cup of coffee in the late afternoon can make it hard for you to fall asleep at night. Nicotine is also a stimulant, often causing smokers to sleep only very lightly. In addition, smokers often wake up too early in the morning because of nicotine withdrawal.
4. Avoid alcoholic drinks before bed. Having a nightcap or alcoholic beverage before sleep may help you relax, but heavy use robs you of REM sleep, keeping you in the lighter stages of sleep. Heavy alcohol ingestion also may contribute to impairment in breathing at night. You also tend to wake up in the middle of the night when the effects of the alcohol have worn off.
5. Avoid large meals and beverages late at night. A light snack is okay, but a large meal can cause indigestion, which interferes with sleep.

---

*Reprinted from *NIH Medline Plus* (Internet). Bethesda, MD: National Library of Medicine (US); summer 2012. Tips for Getting a Good Night's Sleep. Available from https://www.nlm.nih.gov/medlineplus/magazine/issues/summer12/articles/summer12pg20.html.

Drinking too many fluids at night can cause frequent awakenings to urinate.

6. If possible, avoid medicines that delay or disrupt your sleep. Some commonly prescribed heart, blood pressure, or asthma medications, as well as some over-the-counter and herbal remedies for coughs, colds, or allergies, can disrupt sleep patterns. If you have trouble sleeping, talk to your health care provider or pharmacist to see whether any drugs you're taking might be contributing to your insomnia and ask whether they can be taken at other times during the day or early in the evening.
7. Don't take naps after 3 p.m. Naps can help make up for lost sleep, but late afternoon naps can make it harder to fall asleep at night.
8. Relax before bed. Don't overschedule your day so that no time is left for unwinding. A relaxing activity, such as reading or listening to music, should be part of your bedtime ritual.
9. Take a hot bath before bed. The drop in body temperature after getting out of the bath may help you feel sleepy, and the bath can help you relax and slow down so you're more ready to sleep.
10. Dark bedroom, cool bedroom, gadget-free bedroom. Get rid of anything in your bedroom that might distract you from sleep, such as noises, bright lights, an uncomfortable bed, or warm temperatures. You sleep better if the temperature in the room is kept on the cool side. A TV, cell phone, or computer in the bedroom can be a distraction and deprive you of needed sleep. Having a comfortable mattress and pillow can help promote a good night's sleep. Individuals who have insomnia often watch the clock. Turn the clock's face out of view so you don't worry about the time while trying to fall asleep.
11. Have the right sunlight exposure. Daylight is key to regulating daily sleep patterns. Try to get outside in natural sunlight for at least thirty minutes each day. If possible, wake up with the sun or use very bright lights in the morning. Sleep experts recommend that, if you have problems falling asleep, you should get an hour of exposure to morning sunlight and turn down the lights before bedtime.
12. Don't lie in bed awake. If you find yourself still awake after staying in bed for more than twenty minutes or if you are starting to feel anxious or worried, get up and do some relaxing activity until you feel sleepy. The anxiety of not being able to sleep can make it harder to fall asleep.

Illustration Permissions

Figures were provided courtesy of the author except for the following.

Fig. 3. Modified from Noever, R., J. Cronise, and R. A. Relwani. 1995. *Using spider-web patterns to determine toxicity.* NASA Tech Briefs 19(4):82.

Fig. 9. Modified from https://www.ncbi.nlm.nih.gov/pmc/articles/PMC2767184/figure/F1/.

Fig. 10. Modified from http://journals.lww.com/pedorthopaedics/Abstract/2014/03000/Chronic_Lack_of_Sleep_is_Associated_With_Increased.1.aspx.

Fig. 11. Modified from http://www.cbssports.com/nba/news/in-multi-billion-dollar-business-of-nba-sleep-is-the-biggest-debt/. Source: https://jawbone.com/blog/mvp-andre-iguodala-improved-game/.

Fig. 12. Modified from https://www.aaafoundation.org/sites/default/files/AcuteSleepDeprivationCrashRisk.pdf.

Fig. 15. Modified from http://bmjopen.bmj.com/content/2/1/e000850.full.

Fig. 16. Modified from http://www.rand.org/content/dam/rand/pubs/research_reports/RR1700/RR1791/RAND_RR1791.pdf.

Acknowledgments

The staggering devotion of my fellow sleep scientists in the field, and that of the students in my own laboratory, made this book possible. Without their heroic research efforts, it would have been a very thin, uninformative text. Yet scientists and young researchers are only half of the facilitating equation when it comes to discovery. The invaluable and willing participation of research subjects and patients allows fundamental scientific breakthroughs to be uncovered. I offer my deepest gratitude to all of these individuals. Thank you.

Three other entities were instrumental in bringing this book to life. First, my inimitable publisher, Scribner, who believed in this book and its lofty mission to change society. Second, my deftly skilled, inspiring, and deeply committed editors, Shannon Welch and Kathryn Belden. Third, my spectacular agent, sage writing mentor, and ever-present literary guiding light, Tina Bennett. My only hope is that this book represents a worthy match for all you have given to me, and it.

Index